BROKEN ARROW
OF TIME

BROKEN ARROW
OF TIME

RETHINKING THE REVOLUTION IN
MODERN PHYSICS

James Galen Bloyd

Writers Club Press
San Jose New York Lincoln Shanghai

Broken Arrow Of Time
Rethinking The Revolution In Modern Physics

Writers Club Press
an imprint of iUniverse.com, Inc.

For information address:
iUniverse.com, Inc.
5220 S 16th, Ste. 200
Lincoln, NE 68512
www.iuniverse.com

ISBN: 0-595-17874-X

Printed in the United States of America

In memory of my sister, Marissa, who loved me as much as I loved her.

Contents

1

What To Say

Plenty of books have been written on the arrow of time, quantum mechanics, the mind, free will, and so forth. A few authors have had the temerity to handle several of these topics in one book. I have ventured beyond the realistic boundaries of my competence because I see the effort should be made to better understand the universe and our place in it. Admittedly, I expect to be more or less wrong in places and accept the intellectual community to be the final judge on these matters. Nevertheless, this does not worry me. More important is the goal to supply plausible alternative accounts to liven up the debates. This bouncing of ideas and disagreement is central to the progress of science. If there was ever too great a consensus, the process would stagnate and we would be nearing an end, closing the book on science.

Then again, in the public and when we publish, we should also be very careful of what we say in the name of science. The speculation of a scientist is still speculation, and speculation shared by the majority of scientists is still speculation as well. Unfortunately, this is not how the history of science always records things. Sometimes prominent figures earn the popular support of their contemporaries and together they become the authorities, writing our textbooks and teaching our students. Indeed, the scientific community, political and bureaucratic in its own peculiar way, is not immune to certain outbreaks of pseudo-science passing as legitimate scientific knowledge.

Progress in science is often hindered by the tendency to be biased by our perceptions. We often block ourselves with our human centered perspectives. Great advances in our understandings of the natural world have happened when individuals looked beyond archaic preconceptions in the quest for nature's more subtle secrets. Copernicus discerned, against tradition, that the earth is not the center of the universe, Galileo held observation above introspection, and Darwin questioned scripture then learned a natural story of life. These heroes of science are remembered for expanding our world views. When decentralizing ourselves from the world, science is at its best.

Although scientists agree on the necessity for objectivism in natural philosophy, we often fail to incorporate this principle in theory. We are, after all, only human. Certainly, there remains much work to be done in cleaning up our persistent intuitions by sorting out when they are helpful and when they have become obstacles in scientific research.

A particularly widespread human centered view continues to dominate in the scientific enterprise of Physics. The bias is in how physicists typically describe mechanics, both classical and quantum, in terms of an arrow of time. That physicists have continued pragmatically with this temporal bias is not surprising given that the tradition of using laws plus initial conditions have sufficed to describe much of the physical world as we are aware of it. Swinging pendulums and falling apples are phenomena which can be adequately described at a mechanical level where such processes happen over a measure of time. Yet as modern physics ventures more into the unimaginable realms of the very big and the very small, theories of relativistic and quantum processes begin to expose the boundary as to what constitutes a physical event. Experimental philosophers of today have reached one of our greatest challenges in the attempt to comprehend the profound questions raised in modern cosmology and micro-mechanics. It is imperative that scientists begin to seriously answer what time is, and in so doing we must be prepared for another conceptual shift from anthropocentric ideology.

In physical theory we are careful in sifting the rules that are psychological in origin from those existing independently of us. Sometimes this is far from being an easy task. On the subject of time's arrow, this is especially difficult. Throughout recorded history, humankind has sought to answer the perennial question of what is time. We invariably experience time as directional and moving. What could be more obvious to each of us that time flows like an ever-pushing river into the unknown and that we are its progeny heading into the open future of possibilities? That seems undeniable. The mystery lies beneath the perception. What in the world can time itself be? How does time seemingly flow from the material world? That is the mystery.

Understanding the psychology of temporal flow remains to be a problem, particularly for physicists who do not specialize in the science or philosophy of the mind. Often physicists will audaciously derive the psychological sense of time directly from the mysterious arrow of time, though how precisely the former emerges from the latter is not made clear in the literature. Even when keeping within the subject of their own studies, speculating on exactly how an arrow of time can emerge from physics, obscurity has been the norm. Most often directionality is presumed without a proper argument. Various luminaries who have formulated rough drafts for describing a flow of time, have done so by importing temporally asymmetric assumptions not justified by the underlying physics or by relying on other rough drafts which, in turn, contain inserted unsupported asymmetries or themselves rely on further drafts. It is not uncommon to find two or more interdependent hypotheses on time's arrow. For example, irreversible time will be linked with an irreversible process which presupposes irreversible time. It may be the case that the concept of an arrow of time remains in obscurity because it is inherently flawed. Once we rigorously examine the omni-mysterious arrow, we may see that temporal flow is a kind of illusion and carries with it a hefty amount of unnecessary metaphysical baggage.

A few scientists and philosophers have insisted that time does not flow. They argue that we certainly feel time passing but only as an illusion of orientation in space. After all, they ask, how fast does time pass? An answer such as one second per second is not an answer at all. It merely begs the question. To go beyond this apparent tautology, we require a reference point outside of time, something to gauge the flow of time against, but this notion of an absolute time or universal external clock was replaced by the special theory of relativity nearly a hundred years ago. Einstein expressed the situation as "past, present and future are only illusions, however persistent." Time is thus spatialized to be a timescape stretched out in its entirety, like a landscape. The concept is often referred to as block time. Without an external measure for rating the flow of time, we are left with the enigma how we can say that time, as a thing in itself, flows.

Furthermore, of all the fundamental principles in physics, there is no basis for the popular assumption that time objectively flows in a unique direction like some kind of cosmic river. In truth, there is only provisional reason for believing time flows at all. The physical laws of nature appear to be what is called time reversal invariant. This means, that non-statistical mechanical laws are indiscriminate between the past and future directions of time so that each and every event which happens in the usual forward direction of time follows our exact same laws of physics were it to happen in reverse. Nowhere in our non-statistical laws of physics is there anything that says otherwise. Consequently, finding an objective arrow of time has been an elusive goal amid our temporally symmetric physics where all processes are reversible in principle. Nevertheless, block time is not taken very seriously and scientific practice persistently describes phenomena, including those at the edge of physics, in terms of the temporal direction of commonsense experience, fashioning models of causality that act exclusively in one direction, rarely acknowledging the assumptions not required by the experimental data. In quantum theory and cosmology, the flow of time interminably finds its way into the central interpretations.

With a prejudiced theory thus formulated by presupposing time's arrow, we are bound to draw fantastical conclusions. For instance, quantum theory is supplied temporal arrows not required by the underlying physics or the data gathered. This restriction leads the observer to surmise paradoxes, and many attempts to resolve these alleged paradoxes have produced the most bizarre results. All of which have led to immense confusion and will be properly elaborated on further in this book.

A well designed mathematical model generalizes the particulars revealed in physical experiments and disciplinary comparisons, providing useful derivations and predictions applicable to further experiments and comparisons. A well designed mathematics need not literally explain the underlying mechanisms. Yet, overly Platonistic physicists have crossed this line in their conjectures of quantum theory. As will be shown later, quantum theory, to be sure, is compatible with a classical theory of statistics. It is a legitimately applied mathematics for systems in which we cannot know everything so that we can make predictions of single events with limited knowledge. In spite of this, the standard interpretation of modern physics has wholly undermined the rudimentary fact that probability theory is designed for predictions, not explanations. Probability theory, and thus quantum theory and its infamous uncertainty principle, have mistakenly gone from being invented mathematical tools to being the discovered uncertain foundations of the physical world. As will be shown later, generations of modern Platonists have legitimized incomprehension as a science. The studying physicist will often fall into the mainstream interpretation by declaring that indeterminacy is an inherent, inescapable property of the world.

In respect of the empirical methods of science, there is no decisive evidence for accepting the ontology of a flowing time. Perhaps, if we were to suspend belief and begin from the relaxed position that the absolute distinction between past and future are mere artifacts of the human perspective on the world and not of the world itself, then a coherent logic might emerge in modern physics. Nothing radical nor revolutionary need be brought into the discussion. Quite the contrary, the reader will find that one of the messages of this book is that what has been considered revolutionary about modern science is not so revolutionary. By (re)introducing old and new topics into the discourse, the hope is to expose some of the mystery about time with the bonus of achieving a more consistent world view for all of physics.

Absent from this dialogue are the sophisticated arguments preferred by studying philosophers. Rigorous treatments of this sort, no matter how sound and valid, either never reach the non-philosopher or they confuse instead of enrich the reader because of the level of difficulty. To avoid this type of discrimination, the line of reasoning is kept simple and technicalities are kept to a minimum. Fortunately, little substance is lost. By conveniently streamlining, many aspects of the argument are included that would otherwise demand their own books.

Note: The overall argument of this book is based on existing evidence and the resulting scientific models of today. When in it comes to specifics, I bow to the experts in their respective communities. Thus, for evidence and mathematical models, I rely on what is currently accepted by these communities. Allow me to stress the fact that nowhere in this book do I mean to replace or modify any of the theories or laws of science.

Along the way, ideas are employed from various specialties and departments of research, and obviously my knowledge can not be thorough in every topic. No individual fully comprehends all that is covered in the following pages. As many other academic writers have pointed out, we live in the age of the specialist. I am aware of this weakness for an interdisciplinary approach and accept it as a natural consequence. A human endeavor of this magnitude runs the risk of involving errors, but if it were to be precisely correct then it would also be simply boring. To dare to go beyond the mundane has its price as well as its reward.

The plot ahead revolves around the mystery of time. The theme is demystification. In pursuit of a noncircular ontology of time, its essence behind perception must be unveiled. In the answer of what accounts for the flow of time, we want flowing time to vanish into something alien, and only then will the description be satisfactory. It is a funny predicament. Why is it funny? Because it makes us laugh. Well, why does it make us laugh? Because it is funny. You see, to avoid begging the question and to explore the how of something, the magic must be stomped out. Similarly, we do not explain something such as consciousness using conscious-like parts, for example; the Cartesian theater or the soul, because then you have merely repositioned the crux of the problem. The same is true for time's arrow. Replacing the flow of time with an atemporal description will require stepping outside of our sense of time. For many readers it may seem as if I am explaining time away. If later you do feel my conclusions are qualitatively empty, remember that sense and essence are not the same kind of thing and the goal here is to peel back the layer of sensory perception to reveal a more objective picture.

It is difficult to make sense of answers when we do not understand the questions. Once we firmly grasp the problems being addressed and fully understand the difficulties, we might say that the solutions have a way of presenting themselves. Without a doubt, the mystery of time is a slippery problem. It would not help you if I jumped into dismantle-mode now and supplied you with a definition. The mystery, aside from any answer, is largely an ill-defined problem in need of clarification. Therefore, the answer will be suspended until...sometime later. Reading on, you will come across it.

There is a rhyme and reason to the layout of this book. The first three chapters set the stage. No arguments are given quite yet. Chapter one begins with a historical survey of modern science but with a twist. The tell has been told many times over, however, the way it is presented here foreshadows the central arguments of later chapters. I also supply an example in chapter two of the seemingly paradoxical features of modern physics eluded to in chapter one. The renown double-slit experiment which epitomizes quantum weirdness is shown in its full glory. Then after facing the vexing questions of the first two chapters, the reader is led through the third chapter to wonder whether or not any of this matters, the answer of which is left to you, the reader.

From chapter four on, it is a game of tackling the major issues and gaining philosophical ground all the while forestalling the big conclusions until they are more obvious. For protection against the spell of temporal vertigo, in chapters four and five we momentarily diverge from the field of physics into other areas of science and philosophy where the psychology of time is an appropriate topic and more properly dealt with. The second of these two chapters sheds light on problems of free will in the context of block time and determinism.

Between the philosophy of earlier chapters and the physics of later chapters, chapter six is a tour of perhaps the least understood and most counter-intuitive theme of this book. To appreciate the conceptual difficulties of modern physics, it is necessary to have a handle on probability theory. This is not easy, but the chapter begins with some simple illustrations from day to day life and hopefully communicates the essential points.

Thus equipped, we return to physics in chapter seven, starting with relativity and relational models where time is conceptualized as a curve through a spatialized universe. Misconceptions about reversed and frozen time are weeded out, along with other fictional implications of block time, dealing with objections which would otherwise keep popping up in the following chapters. In chapter eight, we move on to a central theme in modern theory, that is; Heisenberg's uncertainty principle, disarming it of its ontological indeterminism by way of some little-known simplifications of what the principle does and does not mean for the philosophy of science. Though the clarifications are carried out with minimal references to any arrows of time, this chapter is essential in preparing us for the conceptual shifts to come. Moving on to chapter nine, we find that the physics community is not immune to myths of its own peculiar kind. We uncover a door to possible explanations of quantum mechanics previously thought to be impossible and clear the path we will be walking on in the next chapter.

Chapter ten is the climax of the story. It claims to have done what many have said could not be done. It supplies an interpretation of quantum theory that makes sense. Assuming all the compelling objections were met in previous chapters, chapter ten provides a model of quantum theory which is not only nonparadoxical, it is one we can adequately visualize.

The remaining chapters apply the same sort of reasoning argued throughout this book to other areas of physics. Refraction, entropy, cosmology, as well as the elusive kaon, are all properly dealt with. Overall, the goal is to unify all areas of physics under one coherent philosophy. The concluding chapter spells out this unity.

One final note: Issues in the philosophy of language are intentionally stepped over. It has been my experience to find questions in the theory of language muddled within their own semantics, stirring up more problems than they settle. The theories are either accurate and trite or profoundly vague. Hence, this book continues in the old tradition of utilizing language as a tool of thought without too much reflection. Hopefully the reader will come to agree that the following discourse does fine without deconstructionism, post-structuralism, or semiotics in general.

2

The Empiricists' New Clothes

Quantum theory emerged in the early part of this century. A growing body of experimental data from microphysics could not be modeled by previous physics, so with creative brilliance and little understanding of what their mathematics meant or why it worked, physicists of the time devised mathematical methods for correlating experimental data with calculated predictions. The clever techniques continued to expand throughout the following decades to include an increasing number of observables in the realm of microphysics.

In the 1950's and 1960's, the field of quantum mechanics was flooded with more inexplicable data pouring from the new particle accelerators. The technology for smashing atoms and subatomic particles was overwhelming physicists with mountains of information. Dozens of theories were concocted to explain the data, but all without success. Physicists had hit a wall. The attempted calculations contained persistent infinities, leaving the equations unsolvable or yielding what seemed to be impossible answers.

After twenty years of heads banging on the wall, equations such as the central Yang-Mills field equations were found to be renormalizable, that is; the infinities could be canceled out or rearranged until they became harmless. This technique of renormalization was first worked out by Richard Feynman who described his own work as sleight of hand tricks to shuffle around these infinite quantities until they either cancel or are absorbed into quantities that can not be measured. Despite the mathematical trickery, renormalization worked in getting the answers experimenters wanted. It was the only successful procedure and was therefore accepted early on into quantum mechanics. When finally the Yang-Mills field equations were renormalized, the gap between formal theory and experimental data was further lessened.

Today, the set of formalisms for dealing with the microworld is known as quantum theory and is utilized by all physicists. It is an enormous success. Few theories, if any, have been verified to the precision that quantum theory has been measured. Advances continue, with few dry spells, elevating quantum theory to the holy grail of modern scientific achievement.

Alas, not all is well. Six decades after its creation the meaning behind the powerful mathematics remains controversial. There is now as much contention among the experts as there has ever been. One of quantum mechanics founding fathers, Niels Bohr, in his famous words told the world, "If someone says that he can think about quantum physics without becoming dizzy, that shows only that he has not understood anything whatever about it." Richard Feynman, remembered as the greatest physicist of his generation, would lecture to his students about quantum mechanics, "Do not keep saying to yourself, if you can possibly avoid it, 'But how can it be like that?' because you will get down the drain, into a blind alley from which nobody has yet escaped. Nobody knows how it can be like that" (1994, 123). Nobel laureate in physics, Steven Weinberg, tells the story, "A year or so ago, while Philip Candelas and I were waiting for an elevator, our conversation turned to a young theorist who had been quite promising as a graduate student and who had then dropped out of sight. I asked Phil what had interfered with the ex-student's research. Phil shook his head sadly and said, 'He tried to understand quantum mechanics'" (1993, 88). Erwin Schrödinger, another father of quantum mechanics, is often quoted for remarking, "I don't like it, and I'm sorry I ever had anything to do with it."

Why are students of particle physics discouraged from asking the question, "How can it be this way?" The colloquial term is quantum weirdness. Everyone interested in science has now at least heard about the uncertainty principle, the double slit experiment, and Schrödinger's cat, and many of us remember the strangeness having something to do with observers affecting their own measurements and particles being waves and waves being particles and such. Since you are reading this treatise, you likely know more details of the strange happenings in labs of particle physicists and that the really weird stuff of quantum mechanics is not touched by the typical hearsay. Many of the circulated conceptions are either misleading or simply wrong. Then again, who is to say what is right or wrong on a subject that no one seems to understand?

For two millennia intellectuals pursued a knowledge of ultimate reasons. Natural philosophers came to view the world as comprehensible. During the last few centuries, science became a tradition of using simple formalisms of logic to mirror the elegant simplicity found in nature. In modern experimental philosophy, however, we no longer assume a coherent reality. Since the onset of quantum theory, incomprehensibility has been promoted to a science. Physical theories have been replaced by theories of probability. Descriptions of objective reality have become ideologies of the past while probabilistic and linear functions have given rise to an ethos of duality and uncertainty. Modern scientists follow statistical processes of the "real world," moving beyond the "dead" mechanics of classical physics.

This splitting between classical and modern physics was well underway in the 1920's where the story is dramatized in the debates between Albert Einstein and Niels Bohr. Einstein, who first developed a quantum theory using photons and electrons, opposed the notion that a probabilistic theory corresponded completely to physical reality. Einstein maintained that there was more to the world than that represented in quantum theory. Niels Bohr heatedly debated with Einstein on this issue. Quantum theory operates on irreducible statistical correlations which Bohr took to completely represent the physical world. He was adamantly opposed to there being an underlying mechanics beneath the mathematical formalisms of quantum theory. Here was a prominent physicist shrugging off the hope of there being a definite reality, and he had a growing number of followers. The physics community eventually sided with Bohr, opposing the living legend of Einstein.

The view of quantum mechanics developed primarily by Bohr, Heisenberg and others while in Copenhagen is referred to as the Copenhagen interpretation, as coined by Heisenberg in 1955. Originally, the Copenhagen interpretation was overly philosophical. Though the philosophy has been toned down, the Copenhagen interpretation retains much of its metaphysical aspects today in what has come to be known as the standard interpretation. The standard interpretation focuses exclusively on the outcomes of physical measurements while forbidding the practitioner from asking questions about possible underlying mechanisms that may produce the observed effects. It treats the mathematics of quantum theory as a nearly complete description of reality. In plain words, it takes the formalisms literally.

Read this way, characteristics of the standard model such as probability, uncertainty, and duality go all the way down to the very substance of what exists, that in the most fundamental laws of the universe there are odds. So our view of the microworld is fuzzy, not because our lenses are out of focus, but because the world is intrinsically blurry. It is not our ignorance of the internal logic that makes nature appear to have probability in it. Thus the popular rhetoric: "Nature herself does not even know which way the electron is going to go." According to this interpretation, Einstein was wrong in saying that God does not play dice, as he has been so often quoted for.

Furthermore, in the original Copenhagen interpretation, the dice are always rolling until someone looks closely at them, forcing them to reveal definite quantities. The dice seem to stop rolling and to give real answers once an experimenter demands certain questions to be answered by the experiment. In quantum theory the mathematical representation of the rolling dice is called the wavefunction and the calculating of the final answer when the dice stops is referred to as the collapse of the wavefunction. The standard interpretation retains the wavefunction and its collapse and views them as more than mathematical. The quantum wave is thought to be an irreducible potential that is somehow real and its collapse somehow physical. Under the standard interpretation, reality is irreducibly random.

Alongside the christened interpretation and its priestly creators, there have remained the heretics who will not leave well enough alone. These physicists remain focused on paradox and an endless preoccupation with meaning and reality and other such human concerns. Their great martyr, Albert Einstein, was tormented by what he saw as a schism at the very heart of science. His later years became littered with corpses of failed attempts to unify classical and quantum physics. Through the eyes of his community, synthesis became the obsession of a stubborn relic. The new high priest, Niels Bohr, preached that quantum mechanics was a theory of reality which would override classical physics once its loose ends where tied. Few scientist have stuck their necks out with Einstein, a few more have taken the fifth, and more than a few have learned to sing the gospels of quantum theory.

3

Sci-Fi Lab

The one experimental embodiment which captures best quantum weirdness is John Wheeler's version of the double slit experiment. Back in the late 1970's, Wheeler suggested his changes to the usual setup in the double slit experiment, and in the mid-80's his version was tested by two independent groups. The results matched the predictions of quantum mechanics, further verifying quantum weirdness.

The message from the double slit experiment is mind boggling enough in its original form. It first emphasized the purported dualism of light as both a wave (light wave) and a particle (the photon), as generalized in the theory of complementarity. Whenever someone would bump up against a new aspect to quantum weirdness, Richard Feynman liked to say to them, "Remember the experiment with two slits? It's like that." You can study this experiment in any introductory course on quantum physics. If you have studied this experiment again and again and are more familiar with it than you ever wanted to be, then be rest assured that the follow recap will be brief.

The trimmed down setup involves a source of light able to emit one photon at a time, a film plate for detecting the photons, and a barrier set between them which has two slits cut very close together. To carry out the experiment, a photon is emitted, passes through the barrier, and strikes the film plate detector, then this is repeated numerous times for statistical purposes.

When the experiment is performed with one of the two slits closed, all is well. The photon is emitted, shoots through the single slit, and hits the detector. After repeated photon firings, we look at the photo detector to see the pattern of distribution, and the film shows an expected scatter with a centered clump of hits. This is akin to witnessing a gifted marksman at the firing range. When you recall the target paper it shows the same scattered pattern, a distribution typical for material particles and bullets.

Now let us open the other slit, run the experiment numerous times and see what happens. This film reveals a very different picture. Instead of having two overlapping clumps of hits that we might expect, we have an interference pattern. But this only happens with waves of particles, like with sound and ocean waves where waves interfere by adding or subtracting to one another's output. There should not be an interference pattern when particles are fired one at a time. The photons are particles when they hit the film plate. We check our equipment, making sure all is working properly. Yes. Then what is going on? Could a single emitted photon split and travel through both slits, interfere with itself on the other side of the barrier and strike the photo film as a distinct particle?

The conventional explanation involves the wavefunction. The wavefunction is said to pass through the setup and collapse when the photon is detected by the film plate then is described as a point particle. Any attempt to detect which slit a photon travels through also collapses the wavefunction but does so at the location of the slit, and once this is done the photon behaves like a particle through the rest of the experiment leaving no interference pattern on the film.

Wheeler noticed a bizarre consequence in using the wavefunction as a description. To highlight the consequence, he modified the double slit setup into a thought experiment that if carried out and confirmed would support the standard description. In his version, an additional detector is placed behind one of the slits to allow a determination of which slit a photon travels through. When the detector is turned on it will allow the passage of the photon yet will also collapse the wavefunction so there will be no interference pattern on the film plate. Thus the photon can be said to travel through only one or the other slit. He then proposed timing the whole experiment so that the detector would not be turned on until after the photon had passed it. Wheeler took this absurdity further and suggested waiting until the last possible moment before the photon reached the plate at the end of the setup. This is most bizarre. It blatantly contradicts our basic sense of causation because, if you did not notice, the additional detector, the one that collapses the wavefunction and determines whether or not the photon will travel through either one slit or both, is placed in the causal chain of events after the photon has already traveled through either one slit or both slits.

Wheeler's setup is appropriately referred to as the delayed choice experiment and is precisely how the experimenters in the 1980's tested to confirm the wavefunction, verifying that light can behave as if it knows ahead of time what will happen during an experiment. It will go through both slits and interfere with itself if it knows pre-cognitively that it will reach the film plate before any other detection along its path. Or, it will pick only one slit if it knows its path will be checked at some time in the future.

Precognitive light? How can light know anything? But, if it cannot know anything ahead of time, then how does an apparent effect precede its cause? The above experiment exemplifies the quantum violation of local causality, but how can it be this way? Well, that is the forbidden question, is it not?

4

So What?

Questions about quantum mechanics trouble most non-scientists. Certainly, they trouble the fresh students of physics. Yet as a young student it is easy to believe the respectable experts and the intimidating professors telling you how the world behaves and not to question it, otherwise you will end up in "a blind alley from which nobody has yet escaped." Nearly all of those majoring in particle physics, after the conditioning in a long, dense education, get used to the conceptual difficulties. They habituate their thinking in the abstract language of mathematics and in dealing indirectly with observables through advanced technologies. The student-turned-physicist ends up saying nature is just like that and applies the tools of quantum theory to research without sweating over fundamental truth or falsity.

This is summarized nicely by Steven Weinberg: "Most physicists use quantum mechanics every day in their working lives without needing to worry about the problem of its interpretation. Being sensible people with very little time to follow up all the ideas and data in their own specialties and not having to worry about this problem, they do not worry about it. So irrelevant is the philosophy of quantum mechanics to its use, that one begins to suspect that all the deep questions about the meaning of measurement are empty" (1993, 88).

Stephen Hawking, likely the most famous living physicist, proclaims, "I don't demand that a theory correspond to reality because I don't know what it is. Reality is not a quality you can test with litmus paper. All I'm concerned with is that the theory should predict the results of measurements. Quantum theory does this very successfully" (1988, 55).

Overall, the message seems to be: "Quantum mechanics is an algorithm. It works. Use it. And do not worry." This attitude may sound familiar to you. It echoes the current philosophy of instrumentalism as well as similar philosophies in the history of ideas for the last two centuries. During the eighteenth century utilitarianism was born in the study of ethics to help objectify the subject of moral coding. The next hundred years of philosophy produced positivism for its all around applicability in studies. Around the turn of the twentieth century pragmatism helped give respectability to the study of psychology, which later influenced the emergence of behaviorism, both of which aided psychology in becoming more of a suitable scientific endeavor and less of a folk study. Even the world of art reflected this theme in the 1960's with minimalism.

The common thread tying the above philosophies together comes back to Stephen Hawking's humble attitude toward an ontology of quantum theory. Epistemologists recognize this sort of stance as opposing the doctrines which have historically prevailed in the West. Instrumentalism and its sister philosophies are breaths of fresh air after centuries of dominating wisdom. When applied appropriately and to sensible degrees these humble attitudes aid in reducing antique dogma and rhetorical ideology.

Newton was a bit of an instrumentalist. After formulating his equations for computing the motion of objects under the influence of gravity, Newton disclaimed any understanding of the physical mechanism involved, leaving it to future generations to uncover the secret of the force of gravity. His equations worked, but how they worked was a mystery to him and others of his time. Many reputable contemporaries dismissed his science as voodoo physics involving a strange force which reached out across empty space to exert instantaneous influences everywhere and on everything. Not until Minkowski spelled out the implications of relativity did gravity become seen as a geometry of the spacetime continuum.

Weinberg, Hawking, and many others deliberate the idea of instrumentalism and will emphasize the enormous success of using quantum theory as a mathematical tool. This is fine for everyone because physicists do not debate over the utility of quantum theory. Toes get stepped on when Weinberg ponders aloud whether or not "the deep questions about the meaning of measurement are empty," and when Hawking admits he does not "demand that a theory correspond to reality." This degree of instrumentalism provokes the persistently classical physicists who have not stopped questioning the nature of reality. These few scientists take the problems seriously.

Roger Penrose, a professor of mathematics at Oxford, insists that whatever reality may be, one has to explain how one perceives the world to be. He argues that quantum mechanics does not do this and one must incorporate something additional, something not contained in the standard rules (1989). After hearing this from Penrose, you might imagine Einstein nodding his head in usual agreement of this sort of stance and repeating his often cited words, "Physical theories try to form a picture of reality and to establish its connection with the wide world of sense impressions. Thus the only justification for our own mental structures is whether and in what way our theories form such a link."

Is an interpretation of quantum theory important? Should we ask of nature what our mathematics and experimental results could mean for reality? The answer seems to be one of personal choice. Although, if we do decide interrogating quantum theory is important, we should heed the wary experts. The precise mathematical formulas we are dealing with have engendered many philosophical speculations beyond reasonable analysis, some of which are highly anthropocentric. Let us not leave the math behind.

Now and then, the questioned is raised: Is it a purely subjective matter as to whether one interpretation is more real than another? Before answering, it is important to recognize that the term "subjective" has suffered much abuse. There is irony to this. Namely, those of us who seriously deny the ability to be objective are not echoing a wisdom of thought because when we embrace the human condition of bias we must go further and deny absolute bivalence in nature and admit we cannot have a distinctly subjective experience without a balancing objective reality, nor can we be biased by who we are unless we can also be influenced by the forces of being.

If in considering the degree of correlation between a fundamental model in physics and an actuality of nature is a purely subjective matter, then what in the world would be an objective matter? What would be the use in performing experiments or collecting data? If everything is a subjective matter, then the terms subjective and objective lose all meaning, and we would simply create new words in their places. These scalar antonyms are part relative, part real, and very useful in both quantitative and qualitative discussions. Can we not keep these terms for describing degrees of bias? Or must we refrain from such talk because some of us cannot get over the shock in waking up to a world that is not black and white?

5

Deception Of Perception

When reading the arguments in later chapters, many readers will naturally wonder how the spatialized pattern of time accounts for the obvious fact that we experience a flow of time. The short answer is that it does not. Attempting to directly relate one to the other is as futile as explaining romantic love using the Periodic Table. It is a fallacy of omitting essential levels of description. It is also academically improper to pontificate on a subject clearly outside of our focus.

Be that as it may, understanding the phenomenology of temporal flow remains to be a problem for physicists who do not specialize in the science of the mind. As stated earlier, it is not too uncommon for physicists to audaciously derive the psychological sense of time directly from the mysterious arrow of time, though how precisely the former emerges from the latter is not made clear in the literature. In fact, if the two are to be closely associated, there is a good argument to be made for the opposite situation of projection, that is; unquestioned temporal vertigo will mislead us into projecting an arrow of time into the outer world. The arrow of time could simply be an feature of the human tendency to place our existence central to everything else, anthropomorphizing the world around us. The pointing out of this possibility leaves it a difficult task for refuting this proposal and arguing why we should consider there being an objective flow of time.

The above dismissal may not satisfy the skeptical reader and rightly so. By demoting the physics of time from its status as a suitable starting point in a depiction of temporal vertigo, it seems the whole problem has been strategically sidestepped. The original question remains to be answered and nothing has been explained. Therefore, it is time we step out of the field of physics and armchair philosophy and into a field where the situation is better defined, into the experimental study of the mind.

Exhibit A (Geldard and Sherrick, 1972, 299-304): The remarkable cutaneous rabbit experiment. A subject's arm rests on a cushioned table with a mechanical setup consisting of tappers placed at three locations along the arm. Then a series of taps are delivered in rhythm, separated by 50msec to 200msec intervals, beginning with five at the wrist followed by two near the elbow then three final taps on the upper arm. The extraordinary effect is that the subject reports feeling the taps scattered along their arm as if a little rabbit is quickly hopping their way up to the shoulder. The first five taps on the wrist are given before the two at the elbow yet the subject experiences the five taps moving up the arm then the two elbow taps also moving up continuing with the last three hops. How can this be?

If exhibit A was not enough, let us move on to exhibit B (Kolers and Grunau, 1976, 329-335); the color phi experiment. A subject is presented with two small spots, separated by as much as 4 degrees of visual angle which are briefly lit in rapid succession for 150msec each and with a 50msec interval. If the first lit spot is yellow and the second one is blue, then the subject will see only one spot that begins as a flash of a yellow then abruptly moves and simultaneously changes color into green then blue before disappearing. Notably, the experience for the subject is of seeing the green changing of color before the final blue even though the changing of color must had been edited into conscious experience after the blue spot was flashed.

Duplicated experiments like these are well documented. The effect of the hopping rabbit and the spot changing color are not manifestations of linear processing within the mind. The apparent stream of consciousness gives way to intermingling dynamics.

If we recline back into our armchair philosophy and treat the flow of conscious as an ideal psychological arrow as well as assume there is a distinct physical arrow of time, we find that we cannot superimpose these two arrows with a perfect one to one correlation. To illustrate, say we followed the measure of time's arrow in the color phi experiment. At the time when the second spot flashed blue, the psychological arrow would be pulling a U-turn heading back to the time when the yellow spot flashed then pulling another U-turn and coming back again following an event of changing colors that never occurred. Neither arrow can possibly be the function of the other so they are either not directly related or one or both of them must be inaccurate.

In looking for a way past the above problem, one might presume that the objective flow of time is also the result of nonlinear dynamics. This way, time and consciousness flow like a river with many currents and slight back drafts but as a whole forcefully running in one direction. The phenomenology of flow can then be correlated with an imagined physical flow of time. The problem with this method is apparent when we zoom in on the rivers of each. Magnifying the rivers of consciousness and of time, we see that these two rivers result from processes that are more than correlated. They are the same river. The only distinction is categorical in that one takes place mostly inside of the brain and the other outside, but empirically there is only one river, the nonlinear flow of time. But now we have gone full circle. Instead of explaining how the arrow of time amounts to temporal vertigo, we have reduced them both to the laws of physics and are right back to questioning whether time flows at all.

We undeniably experience time as directional and moving. Human awareness is a fleeting awareness consisting of an ever-present now. But this temporal vertigo is a fact of consciousness. As hinted at earlier, it is not clear why we should dig as deep as physics in order to explain this human condition. There is good reason why philosophers of the mind, artificial intelligence experts, biologists, psychologists, and others whose specialty is to study consciousness are not rushing to learn quantum mechanics or relativity to help them understand their subject. It is the same reason political scientists can overlook the functioning biology of politicians while biologists can ignore their country's politics. It is why a specialist in quantum theory is not necessarily a specialist in computer programming, and vise versa. It all has to do with levels of functional analogies.

Think of a wave. What makes a wave a wave? For instance, a sound wave is a compression wave, but what does that mean? Here, imagine you and a friend are each holding an end to a long spring or slinky stretched tight between you. Using a free hand you hook a finger into the spring, pulling it slowly toward you. Once it is nice and taunt you let the spring slip from your finger. What you are then witness to is called a compression wave as you watch a compression move rapidly down the length of the slinky. Now imagine the sound of thunder. When lightening strikes, the bolt is so hot, hotter than the surface of the sun, that the air around the bolt expands with explosive force. This becomes an expanding bubble of compression. When we hear and feel this shockwave, we call it thunder. Both of these are examples of the same type of wave.

So what exactly is the wave in the above examples? Again, what makes the wave a wave? It is not the medium. One wave travels along a metal spring, the other through air. Nor is a wave just its constituent parts. No sooner is something a part of a wave than it is behind the wave. Thunder is the result of air molecules pushing air molecules pushing other air molecules until the air molecules in your vicinity push on you. In the case of the slinky, the metal you let slip away is not the same metal pushing on your friend's hand. All we are left with is an action of a specific character. Thus, the two described waves are functional analogies because they share a specific character.

When we want to understand the essence in a functional analogy, it is not important to find every case where it shows up and reduce it to its inner workings. It may never be beneficial. To understand what makes something a compression wave, we do not need to understand what a neutron is. We may find some sort of indirect relevance, but one discussion is not necessary in fully comprehending the other.

The same is true of consciousness and the phenomenology of time. In understanding the stream of consciousness, reductionism is necessary but only to a practical point. Once we have gone below the most pertinent rules, we no longer find interesting answers to our questions about consciousness. Where the most interesting answers lie are in the processes which are unique to the conscious mind. Below this level we begin to find answers which are no longer particular to problems of the mind and seem to underlie more general phenomena. This is why quantum theory, which underlies all of the physical world, helps us little if at all in comprehending what makes are minds special among rocks, stars, and pillows. Does the chair I sit in have a streaming awareness of the present? Certainly not. So industrial designers, and particle physicists, have no special understanding of the somatic mind, and AI researchers, neurologists, and the like remain the experts within their respective fields. Then again, there is no single field of study that does have a special claim to understanding the arrow of time as a qualia of consciousness. In fact, the question is not that commonly asked how a stream of consciousness emerges from a non-streaming, inanimate world.

There are two sides to understanding the mind. One side is in the explanation and is forever debatable. The other side is the one thing we all know equally well, and we know it with complete certainty, that is;

what it is like to have a mind. Thus, one side is what we call quantitative and objective, the other side is qualitative and subjective. Likely the hardest thing to come to terms with, harder than anything else in this book, is how the two sides meet. Whenever a part of the mind is explained to us, the answer seems empty, as if it were missing some essential ingredient because the explanation does not quite match how it feels to be conscious. It is no wonder that many of us cannot imagine creations of artificial intelligence ever really being self aware and alive as we are when we cannot even imagine how our own mechanical bodies do it. Since we feel like the answers are empty we keep looking for the mysterious ingredient which will bring it all together so that we can say to ourselves, "Oh, yes, now it finally makes sense!"

As a teen, I thought that more science meant greater understanding. I naively believed that the more we progressed science the less we would be a people occupied with mysticism or the paranormal or whatever happened to be the current mythologies. But the situation is hardly that simple, is it? Today, there is already enough sciences of the mind, evolution, physics, and so forth to understand naturally, as opposed to supernaturally, how life generally came about and continues to be. Yet orthodox churches still flourish while new religions keep popping up. So it seems, seeing the world with more clarity and less fantasy requires something aside from progressed knowledge or future science. The information to generally understand the mind/body problem is here. More detailed knowledge is not going to make for a more satisfying answer. More detail could likely tie it up in an abstract or technical language, adding to the confusion.

As a child, I remember playing on a marry-go-round with my friends. Sometimes we would get that thing spinning so fast and for so long that we could not hold on any longer. Our fingers would slip and we would go tumbling. Dizzy as could be and a little bruised we would lift our heads to a world still spinning. Then trying to stand became the game. The challenge was to control our balance against the vertigo, to override with our childhood intellect the deception of perception. What a challenge it was.

What a challenge it is. How do we reconcile the way the world seems to be with the way the world actually is? Comprehending the answer is close to answering the questions surrounding mortality because once "the world stops spinning" you are dead. What makes these questions so difficult to answer, at least in a meaningful way, is their dual nature, both objective and subjective, and how we try to answer in one full swoop both sides at once. We might describe the situation as one of complementarity. How can we comprehend the two at once, that is; how can we be aware of the self at the same time dissolving the self into its parts? The one time in our lives when the self does retreat back into being nothing more than its constituent parts, or no longer distinct from the rest of the world, is when we die, but then we understand nothing.

Likely, the best we can do is play a trick with mirrors. We can momentarily forget about explaining "I" and focus on observing others, see others for what they are biologically, mechanically, and behaviorally. Then intellectually know that these perspectives fit ourselves as well. Or, you may have a better technique of your own for reconciling the objective with the subjective. However you choose to think about it, it is a fun challenge, especially so with time's arrow. Under the unbreakable spell of temporal vertigo, one-hundred percent objectivism becomes one of those impossible goals which we constantly work toward but never quite achieve. We are like children who have fallen off a marry-go-round and can not quite shake off the vertigo so that we may stand firm on our feet.

In a highly popular book about time by a physicist, Paul Davies writes, "At the end of the day I am forced to agree…that we are missing something important from the physics of time and our perception thereof…Yet there is an inner sense of time—a back door—buried deep within human consciousness, intimately associated with our sense of personal identity and our unshakable conviction that the future is still 'open,' capable of being molded by our chosen actions" (1996, 276). In these few words, Davies sums up the antithesis of what is put forth by the book you are reading now. Davies is unable to break the spell of temporal vertigo, and because of this, and this alone, he is inclined to believe that the mysterious arrow of time is real in a way that makes it fundamental to our understanding of both physics and cognitive science. He agrees with other popular physicist writers speaking of a "missing link" between subjective time and objective time and goes on to say that we feel time directly through a "secret back door...deep within our souls." This is an example of the worst kind of honest mistake a scientist can make, and regrettably it is all too common for scientists to base their professional opinions on inner feelings.

6

Will To Be Free

For thousands of years, philosophers have puzzled over the problem of free will. Just how free are we? Are we predestined to do as we do by The Unmoved Mover? Is each one of us marked from birth by the orientation of the stars? Can we map the future by mapping the heavens? Do the forces of nature push and pull on our every behavior until we are left with nothing but the illusion of freedom? If our choices are predetermined by some act of God or force of Nature, then how are we to think of our choices, and ourselves, as free?

Determinism is a philosophical doctrine proposing that every event, mental and physical, has a cause. It states that nothing happens without a causal reason behind it, eliminating the element of chance or contingency, meaning that anything and everything we do is determined by strict cause and effect, every outcome is the only possible result of a chain of events, all is set in stone. As with other types of realism, determinism has gained a bad reputation as being a demoralizing doctrine. Its opponents condemn determinism for what they see as a gloomy philosophy of which undermines personal effort and moral responsibility. Ultimately, they see it as depressing and nihilistic. Indeed, the question arises: What is the point of trying to do good deeds, or the point of anything, for that matter, when everything we do is helplessly unavoidable? The aversion is a strong one. If anything I do is inevitable, then what does it matter what I do? These kinds of questions easily lead us into a fatalistic despair.

Free will has two distinct meanings, the stronger of which being the power or ability of the mind to make choices or decisions completely independent of any tangible influences, such as environment, heredity, or brain chemistry. We will call this absolute free will. This strong form of free will as its problems too. An act of absolute free will, an uncaused act outside the causal chain, how could this be possible? Can the mind have moments of pure spontaneity which are truly independent and beyond any sort of cause? Perhaps, more importantly we should ask ourselves if absolute free will is the type of free will we are really after. Consider this: A moment of absolute freedom is when something occurs within the mind that itself has no history, no leading up to of events, that is; a moment of perfect spontaneity, and this pure moment would be a truly random event with no possible reason behind it, would it not? Is this a will that is free? Having things occur in our minds without histories, without provocation, without reasons does not grant us freedom or much of a will. Uncontrolled spontaneity throws our will to the whims of genuine randomness. We lose. By not following the logical continuity throughout the levels of consciousness from which the will emerges and by instead rooting the will in indeterminism and chance, we distort the idea of free will into being an irrational behavior that is out of control. In order for one to be free to make a decision, one must be in control and be the force that leads to that decision. The irony is a marvel; An absolute free will is ultimately no will at all.

Nevertheless, in trying to save us from the fatalistic despair too often associated to determinism, books upon books have been written finding the root of free will in the indeterminacy principle of quantum theory or hiding free will in the unpredictability of chaos theory or freeing the future with an arrow of time flying into the big unknown. There are also always the sects who continue to free the will via the soul or élan vital. Even if we invoke the mystical concept of a soul, we remain stuck with the same problem; that of explaining whether or not the soul, instead of the mind, is either deterministic or spontaneous. Do we consider a soul as being an influence without itself having an influence? A soul may very well be predestined by The First Cause. We just do not know.

Apparently, in order to have freedom one must move on to a weaker form of free will. A relaxed definition merely says we are conscious deliberators who have the ability to make voluntary decisions, based upon (a) our beliefs, desires, and motivations and (b) the conditions and context within which we make these decisions. This weaker form of free will says nothing about either deterministic processes or spontaneity because it remains comfortably within psychology where the subject belongs. Here, we are quite safe. Free will, as self determination, is a given. We all know we can make whatever decision we want to. Philosophically, the implication of determinism for free will is that our decisions are themselves physical processes and have material reasons behind them. These reasons can be hidden and untraceable as long as they do exist. No thought spontaneously appears out of thin air without any cause, be it another thought or perception or something else. When a certain physicist, here unspecified, pondered on paper, "Suppose we live in a totally deterministic universe and all perception of control is illusory…We could never find out if we are puppets or automatons," he missed the point that we are both "puppets" and "automatons," depending on our level of perspective. Going deeper, our thoughts are not affected by cause and effect neuro-processes, but rather they are those very processes. Neurons fire and we feel free.

Over the last three hundred years, as far back as Spinoza, the discussion of free will has been dominated by two schools of thought; compatibilists and incompatibilists. This is an odd alignment given that it is possible to be both a compatibilist and incompatibilist if one recognizes the two definitions of free will. Compatibilist are those who only share a weak definition of free will and see this as compatible with determinism, though they may not be determinists. Incompatibilists are either libertarians who deny determinism or they are hard determinists who deny the absolute freedom of the will. To simplify further discussion, we will skip over the strange boundaries between these camps and plainly speak of practical determinists, who are those viewing free will in the weak sense, as part of consciousness and as part of the cycle of causation, not at the hierarchical top but deeply submerged in the midst of it all.

Now, with the conception of practical determinism in hand, it is time to respond to the negative reaction it often receives. If the future is unalterable, then how can it matter what one chooses to do? Does determinism mean that one's future is determined and that it is futile to act as if one were in control?

First things first, there is nothing inherently pessimistic about determinism. It is an objective model. To read into it and surmise meaning or even moral consequences is an impulse for mysticism we cannot help but to indulge in, however, we ought to acknowledge the leap from ontology to teleology and the fallacy in deriving an exclusive form of the latter from the generalized former. The type of leap we do make is guided by our particular dispositions, attitudes, and what is already held dear to us. Any attempt is a narrow-minded attempt. Thus, the common fatalism described above is not the only derivative of deterministic models. There are others which will be given below. Before going there, let us look at an example of what I am talking about.

Consider the time when we became aware of our genetic heritage and our ancestral link to the rest of life on our planet. This is a type of factual knowledge from which we could interpret numerous meanings. For instance, we could be aghast by our relation to the beasts of this world and thus see ourselves as less noble and more abominable. Indeed, this was and is the reaction we see from many people. Then again, this negative outlook need not be the one we all accept. There are many others. We might see evolutionary theory as a way of connecting ourselves with the rest of life, a way of appreciating our place within this web, or even as a hope that there are others like us out there on similar planets scattered across the universe.

Deterministic outlooks are no different. The nihilistic fatalism described earlier is merely the result of a distaste for determinism common in what is traditionally referred to as westernized societies. In the West, radical individualism is the prevailing attitude, and determinism and individualism do not mix easily. When we who are westernized hear about determinism we react with our sense of self as being an indivisible mode of existence with immutable freedoms and naturally clash this idealization with the doctrine of determinism. That is why the fatalism. That is why the questions centered around personal meaning, such as: "What is the point of what I do? What difference does it make what I decide? Why should I care about what happens since what happens is already determined?" Notice, when we overly focus on questions centered around the self, we quickly lapse into nihilism. Yet, it need not be this way. Nihilism is not essential to the mix of individualism and determinism, not if we extend our philosophy to be more encompassing. With a more inclusive world view, there can be harmony between individualism and determinism. Granted, the harmony does not come easily, but to be able to see it makes all of the steps well worth the effort. Already, most of the work is done in the clarifications made above, and now we only need to follow up with a few more words.

Individuality has often been referred to as an illusion. Certainly, this is an overstatement of the situation, however, it does get the point across. You and I are not indivisible, fundamental things of this world. Not only can we be reduced to our constituent parts or our functional levels of complexity, but there is not even a clear discreteness at any level to define what is or is not you or I. This fuzziness about the self is highlighted most dramatically in cases of brain damage, where the self can literally be of two minds or can be diminished to a near vegetable or can undergo such a personality change as to be unrecognizable by others. The self also changes through time, as we grow up and then grow old. How much is a person the same person they were sixty years ago? Another way to question the definition of the self is to explore its material boundaries. Look at your hand. Is it a part of you? What if it were to be disconnected, cut off, and placed before you? Now, what about the shoe on your foot? As you walk around, how connected to you does it become? How much do you feel the path you walk on through this shoe? From these questions it seems that individuality as a concrete and indivisible idealization is a sort of illusion, but it also seems undeniable that individuals do actually exist. The obvious solution is to then say that the self exists in relation to other things and other selves. How we conceptualize the self will then depend on the setting in which it functions and on the point of view we are utilizing at the time. With this approach, we now have a pliable definition of the self along with a grounded sense of individualism. We have gone from radical individualism to practical individualism.

You have guessed it right. Practical determinism and practical individualism are of the same family. The collective paradigm is logically consistent. Recall, that practical determinists assert that we are conscious deliberators who have the ability to make voluntary decisions, based upon (a) our beliefs, desires, and motivations and (b) the conditions and context within which we makes these decisions. The practical individualists, in this case, defines the "we" simply as the things with beliefs and motivations and the things deliberating. All is well. Or is it? There does seem to be one more thing left unsaid. Yes, and that is, "What is the bright side to determinism?" It was mentioned that nihilism was not an essential interpretation of determinism, that is was one of many possible outlooks. Therefore, this discussion is not complete unless an alternative upside to determinism is provided. Fair enough. One is provided without any further delay. (Keep in mind though, the following does not necessarily reflect the view of the author. It is merely supplied as an alternative.)

A holistic determinist sees chance and accidents as mere artifacts of minds not able to know all there is to know. There are few things which we see as bound to happen while most everything else is uncertain because too much is hidden from view or it happens too fast or the phenomenon is too complicated. But beyond and below this limited human reality there is no such thing as chance. There are no accidents. Nothing is contingent. All is determined. So the holist reflects on this and says that there is a rhyme or reason to everything. The universe is ultimately rational, not because it is conscious but because it is continuous, with causation being its logical link. The holist ponders this further, realizing that he is a thread of this continuity and that he, too, is no accident of this world. If he is superstitious, then he may incorporate this teleology with his mysticism and find comfort in some purpose for his being in knowing that he was not created by chance and that nothing he does is by pure accident. Nonetheless, the non-superstitious will be satisfied in defeating the all too popular claims in the sciences today of his insignificance and of an accidental universe. Purpose or no purpose, the wonder and awe remain when we view the world as deterministic.

7

A Cosmic Skit

The word "probably" applies to our everyday practical decision making. In a universe with regularities, decisions informed by the past are wiser than decisions made at random. That has always been true for humanity, and we would expect non-omniscient organisms such as ourselves to have evolved acute intuitions about consistencies and frequencies in order to have survived as a species.

In the luxury of more modern times, the instinctive "probably" has been codified and refined in the formal ranks of mathematical invention. The early mathematics of probability rose out of the seventeenth century, and percentages came in after the French Revolution with the rest of the metric system, serving tax and interest purposes. More recent is the amount of input to the formulas: teams for gathering data, record keeping, accumulating archives, and tallying and scaling to yield significant answers.

Probability theory is not exactly a worldly passion by most people's standards. It is an applied mathematics, and not everyone does math for the pure joy of it. It is not uncommon for students to take math courses only to fulfill requirements for other courses of study, and even then they only take the minimum. Though everyone gets the gist of what "probably" means, probability theory can be confusing.

Imagine your new job has mandatory drug testing for employees. You happen to be drug free and you really need this job, so you agree to the test. You hand over your fluid sample with a smile. Then a certified lab analyzes your sample. The results: positive for drugs. This is reported to your place of employment, your health insurance company, the blood banks, whoever wants to know. You have been falsely stigmatized.

How did this happen? Was it a fluke? You demand a retest, they deny you, you pay for one yourself, and again the results come back positive. So now no one will ever believe you. You have become a casualty in the fallacy of statistical determinism.

If an expert claims to have devised a test that is 95% accurate, what can you say about a person whose results come back positive? The truth is you can not say anything until you know more about the group from which the person was selected. Say the person is from a group where 5% of the members use drugs, what then? Based on this very limited information, that is; a 95% accuracy and 5% likelihood of being a drug user, any positive test result has a 50/50 chance of being correct. That is it: 50/50! Second and third test do not necessarily lower the probability because false positives are not truly random. False positives occur because of certain chemical reactions that may be idiosyncratic to an individual's body chemistry or diet or over-the-counter drugs or whatever. This is not purely hypothetical either. This is a real world problem.

Probabilities have an odd way of proving how wrong intuition can be. Sometimes the right answer to an apparently simple problem can confound even mathematicians who specialize in probability theory. Likely the best example is the now classic "Monty Hall Dilemma," originally proposed to Marilyn vos Savant (1996, 5). It is worth mentioning that her correct answer to this probability problem has since been verified by classroom experiments and simulated runs at MIT, Los Alamos

National Laboratory, and other schools across the country. Nonetheless, nearly everyone confronted with the "Monty Hall Dilemma" gets it wrong. Massimo Piatelli-Palmarini stated in *Bostonia* (July/August 1991), "Even Nobel physicists systematically give the wrong answer, and they insist on it and are ready to berate in print those who propose the right answer."

How did vos Savant know the right answer? Besides her having the highest recorded IQ at an astounding 228, her correct solution to the game show indicates that she has one of the most thorough under-standings of one of the most confounding theories. Probability theory when applied appropriately is a powerful mathematical tool of brute force. It is tamed only by understanding it. So what is probability the-ory? How are we to think about it?

In short, probability theory incorporates the relative frequencies of classes of outcomes. In calculating a true probability, the details of indi-vidual events must be summed over and categorized. Statistical data is then given or collected for the classes of outcomes. These classes are the actual mathematical entities assigned true probabilities.

In practical terms, no single coin toss is determined by probability theory. Heads and tails as outcomes are classes of possible events where each event is determined by Newtonian-like mechanics. The single event of tossing a coin follows strict causality, spinning this way, falling that way, bouncing here, stopping there, continuously obeying physical law. As a single event, a particular coin toss occupies a mathematically infinite number of classes, which could include a class of tosses that bounce seven times, a class where the coin lands heads-up with George facing west, or a class of coins that get away and are lost. To calculate probability, we must decide on what are called exhaustive and mutually exclusive outcomes of classes, such as simply heads or tails, thereby ignoring or canceling out all of the possible details in individual events. Otherwise, all we could ever say about the probability of events is that each one is both against impossible odds and a sure thing.

Since probability is calculated for sets of outcomes, many theorists conclude that the probability of an individual event is a contradiction in terms. Often a theorists will say that single event probabilities are utter nonsense or that they should be handled by psychoanalysts, not mathematicians. Think back to the drug test. What were the chances of a false positive when your unique natural body chemistry affects the test results? And consider these questions: What does it mean to say there is a 50/50 chance of your guess being correct when you say the color of a card I hold is red when I can see it is black? What are your chances of dying of cancer when you have cancer but do not know it? When you are walking in the rain, what are the odds that a particular raindrop will land on a particular spot on your head? When your parents met what were the chances of you being born?

You toss a coin, catch it, slap it down on your other hand but have not yet looked at it. It is heads or it is tails. The answer is certainly determined. No matter the statistics, no matter the odds you have calculated in your head, the answer is there, regardless of theory. In our purpose ridden minds playing the odds is a way of life. So the unforeseen becomes the accident, the guess follows chance, and only the obvious certainty is predetermined. However, what we see as an accident is really ignorance of the details. What appears to happen by pure chance is a reflection of our limited knowledge of the world. And when we see it coming, nature is making it easy for us. Probability theory is our mathematical trick of getting past limited observation. It is our answer to a universe which has no perceived purpose, a universe colorblind to human labels like certainty and chance. Nature does not discriminate between the accident and the predestined. We discriminate. Therefore, we find odds. For instance, if in a game of cards you were dealt a royal flush, you would think yourself to be incredibly lucky, presuming you knew how to play poker. If you had never heard of poker and could not properly discriminate between different sets of cards, you would look at your same hand not thinking anything of it. Without knowing the rules, without being able to discriminate, no one hand would be more special than another. All combinations of cards would share exactly the same odds. In effect, there would be no odds.

Ever hear that perhaps we, the people of Earth, are just a cosmic joke, a sideshow, or an afterthought in the big scheme of things? What do you think? What if we could rewind the universe several billions years back and let the universe play itself out again, what do you think would be the probability of life emerging? Is life a necessary product of cosmic evolution, or is it a fluke byproduct? Is it, at least, a likely occurrence? According to many scholars, life is a historical contingency, so unlikely that it only happened once in our universe, here on Earth, and even then only by some bizarre stroke of luck. The prominent evolutionary biologist Stephen J. Gould has stated, "Wind back the tape of life, let it replay, and the chances become vanishingly small that anything like human intelligence would grace the replay." Another distinguished biologist, Ernst Mayr, put it this way, "An evolutionist is impressed by the incredible improbability of life to have evolved." Stephen Weinberg wrote that in the principles of the universe "we will find no special status for life or intelligence." Biology, as well as physics, has become a science of the ad hoc.

So what is the probability of intelligent life? Note, this is an all or nothing question, meaning that intelligence either shows up somewhere and somewhen, even if it is only in one far corner of the universe, or it never appears in any part of the universe. The odds? No one really knows. To calculate the probability would require nothing short of a divine statistical analyses. It could be that we are highly unlikely guests in any world or it could be that life would emerge every single time the universe restarted. Even when the question is focused on the history of Earth, there is no realistic way to confidently answer. Say we did pretend to thoroughly understand the fantastical list of precise conditions and circumstances that had to be perfect from the Big Bang and the dawn of life to the emergence of warm-blooded fury creatures called mammals, knowing the physics, chemistry, and evolution of many star systems well enough to declare that the probability of life emerging in any one solar system was precisely one in a million billion. Does this mean the probability of life emerging in the universe is also one in a million billion? Nope, because if there are a million billion star systems around, then the probability is 1:1. Odds do not get any better than that. Then again, this is all conjecture well beyond any real science.

Life sometimes seems unlikely when we think about the evolution of Earth, however, this may be because we fall into the trap of trying to imagine spontaneous life, spawning from a single marked event or act of creation. What is important to recognize is that the questions of origin deal with a set of a long list of non-special missing links in evolutionary theory. Life did not begin in a single event or as a spark or as a fully functioning life form. The process from non-life to life to more life is continuous. What is not continuous is our knowledge of every single step along the way. We can go back hundreds, thousands, millions, or billions of years and find gaps in our story in every category. Why should the gaps 3.5 billions years ago be any more special than those 3.5 millions years ago? Per contra, life seems more likely when we imagine it as an immensely slow emergence in a long span of a continuous stream of events. How about when we look even further? How strange does life look when we follow the stream of events over billions of years?

Here is another approach to the very same question: When you bump into a table, knocking over and breaking a lamp, you say it was an accident, that is; an accident as in being unintentional. However, when a trees falls in a forest and no one is around, is it an accident? Was it unintentional? How about if it is an earthquake which knocks the lamp off of the table? Do earthquakes accidentally knock things over? "Oops, there goes the neighborhood!"

You spot the absurdity. To say we accidentally evolved or to say the universe accidentally created us assumes that the universe has an intentional stance. Furthermore, if this assumption is correct, then we can continue with the absurdity by saying things like, "We were unplanned," and, "God didn't want us." We can literally bastardize ourselves because we are anthropomorphizing nature.

The above is fairly loose terminology. Meanings are unclear and the discussion smacks of being rhetorical. Certainly the scientists who are referring to us as accidental do mean something else by it, something relating to a nonintentional world. Indeed, they do. Nevertheless, as will be shown below, the difference between saying something was an accident and declaring it happened by chance is not as great as we often think.

Probability theory is a tricky business as it is. Throw in modern physics and the problems of each are compounded. In goes the confusion over the uncertainty principle, von Neumann's error, and Bohr's persuasive personality (the three of which are expanded on in later chapters) and out comes indeterminacy, spooky actions and phantom waves, and a science of incomprehensibility.

Let us look at this marriage between probability and physics up close. Where did probability go from being a mathematical tool to being an intrinsic character of the world? If we study the basic Schrödinger wave formulation of quantum theory, we see that this describes the wavefunction as evolving in a deterministic manner. But the wavefunction is unobservable. Only after its collapse do we have a measurement. Precisely how a wavefunction collapses cannot be concluded, only predicted by using statistical analysis supplemented by past observables. By way of calculating true probabilities, quantum theory must deal with sets of case histories, ignoring or summing over individual case histories. The sets then decohere, meaning that all interference has been properly canceled out and grouped so that the case histories for constructing combined sets of histories do not interfere with each other. In plain English, quantum theory does not predict a single event. Instead, it predicts possible outcomes and tells us how likely they are.

Radioactive decay is a straightforward example. The pioneers of nuclear physics found statistical data in the early days of uncovering radioactivity. The rate at which any radioactive substance decays increases in direct proportion to its bulk. To decay means its atoms are jiggling until pieces of their nuclei are emitted. All the atoms in a quantity of radioactive material have exactly the same chance of decaying, so the rate of decaying atoms is in strict proportion to the amount of radioactive substance. In accordance with quantum theory, there is no way to tell when any one particular atom will decay. The best we can do is predict how many atoms will decay in a certain time period. That is it. Not surprisingly, physicists have naively jumped on this irreducible character of radioactivity, reading into the statistics as being proof of indeterminacy. The standard interpretation says there is no physical reason why one atom or another decays, and when a particular atom does decay, there is no physical cause. Absolutely none.

What we forget or have never learned is that a statistical theory is not designed to determine individual events or when any one atom will decay. Thus, the fact that the statistics of radioactivity are not applicable to single atoms is trivial. By the standards of classical physics, radioactive decay is non-revolutionary.

It is important to recognize that determinism and unpredictability are not mutually exclusive terms. In order for something to be a hundred percent predictable, we would have to make a perfect measurement of that something's initial conditions, and for a perfect measurement we would have to have infinitely precise measuring tools. This is simply impossible, by any standards, meaning that whether or not the universe is deterministic, the universe will always be unpredictable.

8

Neo-Classical Physics

There is a certain power in using metaphor. When we come to understand something new, often it is through familiarity. Thousands of years ago a humble Greek philosopher by the name of Socrates developed an analogy for wisdom. He told the story of prisoners held in a cave. They are bound, unable to move, and to spend their lives staring ahead at the cave walls. Puppets and figures are paraded in front of a flickering fire casting shadows on the walls and only the shadows are seen by the prisoners. Knowing nothing but the shadows, the prisoners come to believe the flat world is all there is and devise theories of motion and morphology based on their dark universe. A wise prisoner is able to look back at the fire and puppets then break away from their shackles eventually leaving the cave.

Two thousand years later, a similar story was told, this time in London, 1884, by a man named Edwin Abbot. As with Socrates, Abbot's analogy had many messages woven into it, both political and metaphysical. Abbot's novel is a tell of Flatland, a two-dimensional world whose inhabitants are lines, squares, polygons, and circles. In its climatic scene, a three dimensional being, Lord Sphere, peels Mr. Square off of Flatland

and sets him floating in Spaceland. When Mr. Square returns to Flatland his worldview has been forever changed.

Say we believed the Earth was flat and went about surveying an expanse of land. In taking measurements we would notice something peculiar. When we make crude measurements of land, all is well, but as we use increasingly accurate methods we would realize that calculations involving sine, cosine, and tangent give contradicting answers. The more precision, the more the numbers calculated on paper would not correlate with our surveys of the land. Degrees in triangulation techniques would be off and parallel paths would eventually intersect. These answers would keep popping up that were counter-intuitive. Because of our false expectations we would be forced to add what seemed like arbitrary constants and mathematical tricks to our equations to compensate for the gaps between theory and data. Being confronted with the illogical results, we might end up saying, "Oh, nature is just like that, and the problem lies in how our brains are wired and how we've evolved to think so don't even try to make it coherent since it's just one of those things we are too limited to understand." If you read much scientific literature you will notice that, indeed, this seems to be the prevailing attitude on quantum theory in general.

In surveying land, our calculations need to include the curvature of the earth if our results are to be accurate. We cannot ignore two millennia of geometry. When measuring any surface, we can use a distant perspective where the two dimensional realm is curved or wrapped up through a third dimension. Adding this element of curvature then reveals a hidden logic, ringing bells in our heads so we say, now that makes sense! To take the analogy further into the predicament of today's physics, the philosopher at the University of Sydney, Huw Price, says, "If we see the historical process of detection and elimination of anthropocentrism as one of the adoption of progressively more detached standpoints for science, (then) physics has yet to achieve the standpoint required for an understanding of temporal asymmetry. In this case the required standpoint is an atemporal one, a point outside time, a point free of the distortions which stem from the fact that we are creatures in time -truly, then, a view from nowhen" (1996, 21).

Time is not just another spatial dimension. It is time. Then again, none of the dimensions are just another dimension. If you have a one dimensional boring world like a simple line and add another dimension, then this second dimension is not just another dimension. It dramatically enriches the world with an all new dynamics. The same goes for Flatland, and I mean flat with absolutely no thickness. When you turn Flatland into a pop-up world....well, the change is miraculous. Then when you take this three dimensional world and blow it up through a fourth dimension, that of time, and into a spectacular and complex universe, then you have got yourself more than just a few dimensions. So time is not just another spatial dimension, but then really it is.

Neither is time simply an illusion, no more than space. We truly are creatures in timespace, and as much as anything can be real, time is real. Still, this is not to say that the *arrow* of time is more than a subjective sense of vertigo. Temporal vertigo is analogous with the regular **vertigo** experienced when you step off a marry-go-round after spinning for **too** long. You step off but the world keeps on spinning at a dizzying **speed,** and this is a strong hallucination which you only know is false **because** of your reasoning. Otherwise, the motion is so real it knocks **you on** your butt.

Time is a crucial concept in physical science. Experimentation is founded on the ordering of events in causality. Physical laws are expressed in the form of differential equations with respect to time, thereby incorporating the notion that they must explain what happens in terms of preceding instances. In this accordance, physicists solve equations on the basis of initial situations. We decide beforehand what is to be considered the initial state and the resulting state. To solve mechanical formulas, quantum or classical, we first assume a temporal direction, and by doing so we create a particular arrow of time.

This explicit character of physical theory omits the current laws of physics from being on the list of possible explanations for an arrow of time. In order to scientifically account for an elusive arrow of time, we require something fundamental to that arrow, and modern physics certainly is not fundamental. Richard Feynman would admit to his students, "In all the laws of physics there does not seem to be any distinction between the past and the future. The moving picture should work the same going both ways, and the physicist who looks at it should not laugh" (1994, 103).

As a case in point, let us look at the behavior of moving electrically charged particles in interaction with an electromagnetic field. The Maxwell-Lorentz equations describe this behavior, and since these are partial differential equations, their solutions require the specification of boundary conditions and initial conditions. Traditionally, the parameters are met by assuming a direction of time from past to future, in which the calculated answer is then called the retarded solution. The Maxwell-Lorentz equations do not inherently supply the temporal direction so for every retarded solution there is a mirroring result called the advanced solution, which correctly describes the depicted event in reversed causality.

As Feynman clearly stated, there is no apparent arrow of time at the core of physical laws. Temporal direction must be inserted by the physicist. Regardless of this fact, physicists, including Feynman, intuit a flow of time anyhow, constructing theories based on this temporal bias. As a result, there has been much confusion as to what something like the overlooked advanced solution might mean for objective reality. Typically the advanced solution is treated as distinct from the retarded solution, involving separate interacting histories each aligned with arrows of time flying in opposite directions.

The confusion possibly stems from how physicists visualize hypothetical problems dealing with time. The phrases "reversed time" and "reversing time's arrow" are often interchanged though they are not analogous. The prominent physicist, Paul Davies, in his recent book *About Time*, shows ambivalence on the subject by being skeptical of time's arrow in the same chapter where he imagines reversed time involving "broken eggs reassembling themselves as if by a miracle, water running uphill, snow unmelting into snowmen,...we would think in reverse...our mental activity would be inverted too" (1996, 222).

In Philip K. Dick's fictional *Counter-Clock World*, the direction of time is flipped in 1986. Time is reversed, creating a phase in which entropy decreases instead of increases. During this period coffee separates from cream, billiard balls roll into neat triangles, glass fragments come together forming glasses precariously perched on the edge of tables, and so on.

These misconceptions of time run far and wide. To claim that reversed time means any sort of change to the universe and would result in inexplicable events is going beyond sober analysis. What Davies' and Dick's stories are referring to is a reorientation of the obscure arrow of time, not simply the notion of measured time. They are assuming there is an arrow of time then imagining reversing its direction. This distinction is critical in understanding what reversed time actually entails.

Both the contestants and supporters of an objective arrow of time often fail to distinguish between time and time's arrow. One common mistake to avoid involves the popular analogy with a movie film. In this relation, the universe is like a movie with past, present, and future all already existing on film from a perspective outside of this movie, from the perspective of the one holding the film in their hands. They say running the film on a movie player is analogous to how time brings life to our universe. The flow of time is supposedly like playing the movie. But the analogy is merely repositioning the lump in the carpet or replacing one mystery with an equal one. Instead of explaining time's arrow, they simply introduce new terminology so we are left with the problem of explaining what playing the movie of the universe means when we have no idea what sort of thing "playing" is.

Furthermore, if the flow of time for the universe is like playing a movie, what happens when the playback is sped up? In this hypothetical scenario, we in the universe, like the characters in a film, would not notice a difference between play and fast-forward because everything in the universe would also speed up. All motions and relative frames of reference would be equally affected, and as creatures within the fabric of the universe, we could not possibly notice. Play the universe faster or slower, it makes no difference.

What if the film is paused, stopped, or never played at all? What if the video cassette just sits in the VCR and no one ever presses play? None if this could possibly be noticed by the characters in the film, nor by us in the universe. Play the universe, pause it for a moment, play some more, fast-forward, and we would never notice these altering states of affairs because all frames of reference would be equally influenced. Play again, slow down, stop, rewind. . . Yes, even rewind would not be noticed. Rewind the universe and we could not possibly notice as long as everything is being equally rewound. Play, stop, and rewind only have meaning when there is an actual external clock to measure against. If we invent this external clock to explain an internal clock of the universe, then we are no closer to our goal of explaining exactly how time flows. An explanatory model of time's arrow must create the flow of time but do so without already assuming or using the very dynamic we are trying to prove. For example, we could relate our universe to a movie, but in doing so we must not require the additional element of playing the movie. We need to dispense with any flowing spatial-temporal framework. The video cassette of the universe must be allowed to sit on a table never to be played, and let that be enough. Otherwise, the best we could ever do in explaining the character of time would be saying, "It just is that way," or, "Then a miracle occurs" somewhere in our argument.

You have heard someone say, "Imagine if time freezes for a moment then starts again. We would not notice because we would have been frozen too." The imagery presupposes an arrow of time and assumes that psychological temporal vertigo cannot proceed if time does not flow. The presumption is reminiscent of the Zombie Dilemma where we ask the perplexing question: What would happen to a person in perfectly good health if suddenly their consciousness somehow faded and left their body? Could they be a walking, talking zombie? Would they even notice? Could someone be born a zombie? How could you tell whether or not someone else was actually conscious or just a zombie?

These puzzles persist only when we hold onto the obscure notion of consciousness being purely intangible, free to exist separate from the body. On the other hand, if we discard Dualism and look to see how the mind might be the action of the body and find how the two are interconnected, then the Zombie Dilemma turns out not be a dilemma at all but a failure to imagine how non-conscious parts can relate to a conscious whole.

Perhaps, the same can be said for frozen time. If we discard the dualism between time and time's arrow and look to see how a flow of time might be the result of measured time and find how the two are interconnected, then the freezing time imagery could be simply a failure to imagine how non-flowing time can relate to the phenomenology of flowing time. This way both a soulless body and a non-flowing universe could account for consciousness and temporal vertigo without conjuring any spirits or working any miracles.

If the idea of time being a spatial dimension is to be taken seriously, then there ought to be a good answer to the question: How is it that we are free to move through space in any direction but we are limited temporally in that we can only go forward? In the usual dimensions of space we can go up or down and left or right, but when it comes to traveling through time we have no choice in the direction. Why is this? The answer is in this century's most quoted equation.

E=mc^2 means that mass and energy are two elements of the same principle. It also tells you that from an imagined point of reference everything in this universe shares a constant value for a particular velocity vector. You and I and the sun and whatever else are all progressing at the same rate through spacetime. The common notion of relativity applies when you distinguish how much energy is being used by a frame of reference to move through space as opposed to how much energy through time, but the total of that ratio is constant and ubiquitous. Another way to illustrate this is to graph it out. Roughly, the x-axis will represent the change in space, and the y-axis will represent the change in time. If you want to represent a velocity vector for photons of light, you can draw an arrow parallel to the x-axis. If you wish to represent yourself, you can etch another arrow nearly perpendicular to the first. The important rule to follow when filling your graph with vectors, according to relativity, is that all vectors must have equal magnitudes, that is; the length of the vectors for different objects are all the same while the slopes, or ratios, vary. In the real world, this means the total amount of change in spacetime is the same for everything. The point of reference being the one you used to draw the graph.

So even as you sit there reading this, you are traveling at the velocity of light with your vector nearly horizontal along the y-axis of time. This explains why light, having its vector vertical on the x-axis of space with all of its energy used for space travel and none temporally, holds the world's record in spatial velocity. With this in mind, you can see roughly how the velocity of light is more of a constant than it is a limit in speed.

Another recognized lesson from relativity begins with the question: What would we see if we were traveling at the spatial velocity of light? The answer is that from this special vantage point, the universe is collapsed. Notice the smallest of energies like those of light are timeless in this respect. If we, too, could travel only through space and not time, we would not perceive ourselves to be traveling at all. Our points of departure and of arrival would be the same point since no time would pass for us between the two. From this viewpoint, dimensions would be collapsed so past, present, and future would be one.

Now, let us turn this around, following other equally logical consequences. Instead of imagining temporal reality as collapsed to a point, visualize the light as expanded throughout spacetime. Light can be seen as not traveling but simply already at its destination (absorber), its point of origin (emitter), and every point in-between (quasi-standing wave). From this unique perspective of nowhen, we can see another hint as to why no time passes for a photon. Light is seen as a thread of spacetime because its travels are simply an expanded matter of existence. From nowhen, light is not described as traveling, no more than a railroad track travels across land.

Following this logic further, we can extend the imagery to all light as well as all fundamental energy, expanding them temporally so that the entire history of the universe, past and future, will already exist as a whole. You may recognize this as depicting the universe as being superdeterministic. This is correct. If the universe exist continuous and whole, past and future are connected as one. Thus, the future is determined by the past in as much as the past is determined by the future.

With there being no external nor absolute space or time, the above is a relational model of the universe with respect to relativity, along the vague philosophical lines of block time and a block universe. With this handy, we are now ready to address the question of the universe having a constant velocity, c, represented in the vector magnitude graph. Traditionally, velocity implies a motion relative to something else so when we say the universe and all of its parts have a constant velocity through spacetime, we naturally want to know what this velocity is relative to. But ever since the philosophy of Leibniz, the mathematics of Hendrik Lorentz's transformation equations, and the science of Einstein's special theory of relativity, the spacetime of the universe has been postulated to stand on its own without any pre-existing geometry. In a superdeterministic model where time is spatialized, the constant, c, can be thought of as not a velocity but rather as a curve through space-time. Though the specific values of measurement may be arbitrary, this treatment does not require a background or an absolutely motionless ether to be measured against. Nothing need exist external to the universe as would be the case if the universal constant c were to be thought of as a velocity or rate of change above all else. (It also introduces some fresh answers to the ancient pre-calculus motion paradoxes.)

Note: explicit mathematical treatments of relational models has been performed by those such as J. Barbour, A. Ashtekar, C. Rovelli, and L. Smolin where they formally show how time does not need to be postulated as a part of an external kinematic frame. It can be recovered purely relationally from a timeless geodesic principle in the relative configuration space of the universe.

9

Uncertain Uncertainty

In the late 1920's, Werner Heisenberg formulated a mathematics for indeterminacy within the context of quantum mechanics. When working out the details with his colleague, Bohr convinced Heisenberg that the subject of quantum indeterminacy was more than mathematical in nature, and together they added it as part of their developing Copenhagen interpretation. In 1932 Heisenberg was awarded the Nobel Prize for his uncertainty principle.

We have all heard about the uncertainty principle. Typically, most of us vaguely remember it having to do with disturbing what we measure. In popular media, an example would be trying to take the temperature of water in a small glass using a big thermometer. The thermometer will transfer heat with the water as it reads the temperature, changing the original water temperature. Another crude example would be trying to measure the temperature of your freezer by opening the door and sticking your hand inside, knowing full well that you will flood the freezer with room temperature air.

These careless paraphrases, however, have erroneously associate the uncertainty principle with the fact that making a measurement effects what is being measured. Physicists typically shake their heads at the popular misconceptions. The above problem with a thermometer and a glass of water can be overcome and is not limited by an uncertainty principle. Instruments effecting what is being measured has been and will always be a variable with or without any mathematical uncertainty principles. The popular pseudo-uncertainties are really problems of ingenuity where sought answers can be achieved with the proper methodology.

A genuine uncertainty principle arises when there is a computational link and limit between two variables. Precisely, the increasing of accuracy in measuring one quantity decreases the accuracy with which another quantity may be known. Given that an uncertainty principle is a mathematical equation showing the relationship between two variables, neither of which is the instrument of measurement or what is being measured, there is no analogous principle when you only have one sought variable, such as temperature, when using a thermometer. In quantum theory, the variables being referred to are interdependent properties such as time and space, momentum and position, and energy and matter. For instance, to have complete information for position means no information for momentum, and vice versa. To have some information on position means you can have some information on momentum, but the amounts are directly proportional to one another. More of one means less of the other. Quantum theory implies an exact tradeoff between the two complementary properties.

Einstein, like many other physicist of his day, overlooked the mathematics and argued that uncertainty simply resulted from clumsy deficiencies in experiments. He dreamed up thought experiments which he hoped could in principle measure completely position and momentum, thereby refuting Heisenberg. Each time, Bohr would find a flaw in Einstein's hypothetical argument. Bohr's success in this debate discredited not only Einstein's views on microphysics but classical physics in general and was a pivotal reason why indeterminacy has become the established way of thinking about quantum theory.

The standard interpretation of quantum theory, as evolved from the Copenhagen interpretation, retains the two inter-linked core features, that is; Bohr's complementarity and Heisenberg's uncertainty principle. The standard interpretation treats the uncertainty principle as unique to quantum theory, and thus attributes the mathematical limit as being proof for an intrinsically indeterminate world. So it is argued that within the limits set forth by Heisenberg's uncertainty principle quantum particles do not have things like precise kinetic energies, or, as the saying goes, "An electron does not know both where it is and where it is going." This conventional view frowns upon theories which imply classical notions such as a determinate objective reality. The contemporary reasoning is that the quantum uncertainty principle as a fact of microphysics is incompatible with classical physics or determinism, but, at the same time, necessarily follows from today's standard interpretation of quantum physics.

As mentioned above, the more informed people shake their heads at the public's misconception of the uncertainty principle. In turn, modern physicists shake their heads at Einstein's fumbling. Now I dare say it is time we shake our heads at the physics community and the standard conception of what the uncertainty principle translates into. But before explaining why, allow Professor Lee Smolin another confession from the physics community; Smolin: "I must stress that I do not know why Heisenberg's uncertainty principle is true. Neither, as far as I can tell, does anyone else" (1997, 172).

The uncertainty principle is a key insight into the generally accepted interpretation of quantum theory. Consequently it has received much attention in the scientific enterprise, been the subject of many articles and books, and stood center-stage to the quantum physics vs. classical physics debates. Yet, contrary to conventional wisdom, the uncertainty principle actually has roots in classical phenomena and is easy to understand from a viewpoint outside of quantum physics. It is not peculiar to the Schrödinger equations of quantum theory. Heisenberg's uncertainty relations result from a mathematical quirk found in every linear time-invariant mathematics. All conjugate variables, that is; variables whose product results in units of action, give rise to uncertainty principles.

In electrical engineering, there is an uncertainty for signal processing. The property of Gaussian distributions under Fourier transformations is well known and understood. As an example, consider the representations of fast electrical pulses in time and frequency domains. Such a pulse can be represented either in the time domain as a set of voltages varying continuously as a function of time or in the frequency domain as a continuous set of Fourier components. These representations of such pulses have exactly the Bohr-Heisenberg relationship and exhibit their own precise uncertainty principle. It is not possible for the widths of the graphs to both be made arbitrarily small, a fact which underlies the bandwidth theorem of signal theory. The role of the Gaussian function in providing the minimum uncertainty product is related to the fact that it is not possible for both a function and its transform to decay arbitrarily rapidly at infinity. This electrical engineering property can be observed on an oscilloscope screen in your nearest electronics laboratory. Here is a purely classical phenomenon which exhibits true uncertainty.

In layman's terms, we have uncertainty principles because we are using a system of measurement and style of math which work just fine at the level of our everyday world where we drive cars, play baseball, and sit at the beach watching waves crash to the shore. Here, consider wave/particle duality. Duality is a lingering idea which endures in physics by virtue of our drive to make the microworld analogous to the macroworld. In the everyday macroworld, waves are always an action involving many objects or particles. A particle, such as a dust particle, can be either moving or motionless and can be broken up into many sub-objects. But at the level of the microworld, these concepts fall apart. A wave becomes unobservable and can transverse space without a medium, and a particle gets a wave all to its self, neither of them behaving like their macro-counterparts anymore. They become more abstract and inconceivable. Yet we still use these terms in our descriptions of quantum mechanics as if the microworld is analogous to the macroworld, instead of allowing these terms to dissolve into something fundamental. If we at least used words like pseudo-wave or quasi-particle, then perhaps we would not be so surprised by pseudo-wave/quasi-particle duality and maybe even expect this type of blurring. Duality as a mathematical engineering problem for the quantum world is only counterintuitive when we carry the wave and particle analogies too far. After all, if a quantum entity can have the labels spin, color, and flavor, which they do, then why should momentum and position retain complete Newtonian meaning?

Stephen Hawking's positivism shined when he said about the wave-function, "The unpredictable element comes in only when we try to interpret the wave in terms of the positions and velocities of particles. But maybe that is our mistake: maybe there are no particle positions and velocities, but only waves. It is just that we try to fit the waves to our preconceived ideas of positions and velocities. The resulting mismatch is the cause of the apparent unpredictability" (1988, 173). Professor Bart Kosko of USC describes the mismatch as "the math modeler's dilemma: linear math, nonlinear world" (1993, 108).

The uncertainty principle has a misleading name. It is a clearly expressed equation, and we can be certain of its mathematical form. It is no more a mathematics of uncertainty than relativity is. In fact, its significance is very similar to that of relativity's. They both link measured properties in physics such as both time and space and both energy and matter. These linked properties have since been understood to be interdependent and inseparable. Indeed, what is time without space, energy without matter, or momentum without location? It was by a curious historical accident that the uncertainty principle got its name and relativity got its and that we today do not speak of time as uncertain instead of relative.

Restated, the uncertainty principle is a limit derived from the applied math in microphysics. Thus, indirectly, it follows from experiments. Directly though, it is quantum theory reorganized. Being that it is directly surmised from the mathematics of quantum theory, the principle is not about disturbing what you measure. Nor is it a tenet for an indeterminate world. All that is certain is that Heisenberg's principle is an interesting linear mathematical oddity found in both classical and quantum physics, and points at no essential distinction between the two. It calls for no revolutions in scientific thought, no denial of a determinate world, and no probabilistic ontology.

Heisenberg's uncertainty principle is often interchanged with Bohr's complementarity principle. Like the uncertainty principle, complementarity was a cornerstone of the Copenhagen interpretation and remains an essential feature in today's standard interpretation. But whereas Heisenberg's principle is a precise mathematical derivative of quantum theory, Bohr's principle certainly is not. The latter is a notoriously vague doctrine as far as science should be concerned. When we talk about the difficulties in experimenters disturbing what they measure, it is often the complementarity principle we are actually referring to. In its clearest form, complementarity means that the application of one classical concept precludes the simultaneous application of another classical concept when measuring quantum phenomena. This sounds suspiciously like a rewording of the uncertainty principle, however, it strays from science by being mostly philosophical. Remember, Heisenberg's principle is not about disturbing what we measure but about precisely linked variables of which are being measured. Indeed, complementarity can be seen as an unnecessary philosophical amendment to the uncertainty principle. In fact, the term "complementarity" was first popularized by the philosopher William James around the turn of the century and before Bohr introduced the term into physics. James, one of the fathers of pragmatism and experimental psychology, used the term in a field unrelated to physics, but the general idea was the same, revealing a bivalence between two perspectives while showing how the use of one seems to exclude the other. Likely, it is no coincidence that Bohr later expanded on the usage of complementarity to include examples outside of physics. In a theme remarkably similar to James', Bohr would make analogies between the dualism found in quantum mechanics and the dualism apparent in other studies such as biology, psychology, and cultural studies. At the International Congress of Anthropological and Ethnological Sciences in 1938, he described emotions as standing in a complementarity relationship with perceptions of emotions.

By the 1930's, Bohr had developed a reputation as being nearly a quantum guru and many of the prominent physicists of the time became enthusiastic supporters of complementarity. Of course, there were a few who were hostile toward Bohr's doctrine. Besides Einstein, the British mathematician and physicist Paul Dirac, a Nobel prize winner, did not see any usefulness in Bohr's philosophy. It did not calculate results and made no predictions for new equations. In the United States, the reception was neither supportive nor hostile but more of a lack of interest. For the most part, the growing number of young physicists who were making a name for themselves in the states were more intent on doing experimental physics and did not concern themselves too much with the philosophical aspects of their work. Ironically, they were being pragmatic about quantum mechanics, an attitude which was earlier bolstered by the American William James. In spite of this, complementarity did eventually become established as part of the dominant view in the community of quantum physicists.

Quick Note on Planck Time: Contrary to popular belief, Planck time does *not* represent indivisible time. Planck time is a non-arbitrary unit of measurement derived from the three constants of nature; the speed of light c, the force of gravity G, and the ratio of energy of any quantum of radiation to the vibration frequency of that radiation h. With these three constants, c, G, and h and some algebra one can formulate units of measurement for energy, mass, force, length and time. These are then referred to has Planck units of measurement which are not arbitrary like meters and feet. But none of these are any less indivisible than a meter or foot. As a case in point, the Planck mass is a hundred thousandth of a gram. That's huge in comparison to the actual masses of subatomic particles. More outstanding is the Planck energy which is a million billion times greater than any energy reached in accelerators!

10

Spooky Scientific Voodooism

Einstein objected to quantum mechanics, as developed by Bohm, Heisenberg, Dirac, and others, because of the many implicit features that conflicted with Einstein's understanding of the universe. Over the years he developed a defiant list of objections to the peculiarities of quantum mechanics. At the top of the list was what Einstein called spooky actions at a distance. Einstein's spookiness is now referred to as nonlocality, that is; isolated parts of any system which are out of speed-of-light spatiotemporal contact with other parts of that system retain definite quantum correlations not possible through classical causation but are somehow being enforced superluminally or faster-than-light.

In 1935, Einstein, Boris Podolsky, and Nathan Rosen, published the EPR paper, in which they argued against quantum mechanics because of its nonlocal character. Bohr and his supporters tried to come to grips with the EPR criticisms, and a long battle ensued. During this time, the fight was held on theoretical grounds for no clear cut experiments were performed to determine if nature exhibited nonlocality.

Local theories are often confused with hidden variable theories. Hidden variable theorists seek out hidden mechanics possibly over-looked by quantum theory. One such theorist, David Bohm, worked off of Louis de Broglie's incomplete theory, introducing a local hidden variable theory that would replace orthodox quantum theory with a theoretical structure omitting the paradoxical features referred to by the EPR paper. Physicists, however, paid little attention to his alternative. Bohm's approach was less useful for calculating the behavior of physical systems and his conjured pilot wave seemed no less miraculous than the standard notion of wavefunction. Since it was apparently impossible to resolve the EPR debate by performing an experiment, physicists tended to ignore the controversy over both locality and hidden variables. The EPR objections were considered problems for philosophers and theorists, not working physicists.

Besides, all hidden variable theories were undermined by calculations carried out by John von Neumann in 1932. Von Neumann was considered one of the top mathematical geniuses of his generation. So when he claimed to have formally proven that no hidden variable theory could ever account for quantum phenomena, physicists believed him. For more than three decades, physicists accepted his proof at face value. "The hidden variable theory cannot be true," said Richard Feynman in his 1964 lectures (1994, 140).

In 1966, John S. Bell, another well renown genius, found a fatal flaw in von Neumann's calculations and showed the proof was based on a mathematically false assumption. After thirty-four years of physicists dismissing hidden variable theories, you could say the news came a bit late. Ironically, John Bell formulated an inequality which was popularly mistaken as defeating hidden variable theories. Many years later when Bell's inequalities were experimentally verified in Aspect's 1982 experiment, the results were further misinterpreted as a demonstration that hidden variable theories must be wrong. Nearly tens years later the physicist David Lindley was consequently mislead into saying, "Alain Aspect and his group got the result that many expected: the standard interpretation of quantum mechanics was correct and the hidden-variable idea was not" (1991, 103). A year later Paul Davies and John Gribbin (Gribbin has recently correct himself on this subject) also said, "The Aspect experiment lays to rest Einstein's hope that quantum uncertainty and indeterminism can be traced back to a substratum of hidden forces at work. We have to accept that there is an intrinsic irreducible uncertainty in nature" (1992, 244) and, thus, no hidden variables.

Slowly the misconception is being recognized as physicist realize that Bell's theorem assumed a local, nonspooky theory in its inequality. Bell's inequalities and the resulting experiments demonstrated that any local model whether it be hidden variable or not was incapable of explaining quantum phenomena. Aspect's experiment verified nonlocality. Therefore, nonlocal hidden variable theories may be entirely consistent.

Then again, there is a loophole for locality as well, one not usually discussed in the physics literature. The proof of Bell's inequalities requires a temporally biased assumption known as independence, that is; variables at the time two systems interact are independent of later measurement settings. In plain words, Bell assumed later quantum states do not in any way affect earlier quantum states, meaning there is absolutely no backward causation at the quantum level. So if we do allow observables to depend on past and future measurement settings, then a local theory can explain quantum phenomena. Interestingly, this principle for such a local theory defies no fundamental laws of physics. In fact, the mechanics for this type of super-causality already exists in elementary physics. As far as the laws of physics are time symmetric, they can be used to formulate causation in either direction of time. Thus, taking this idea onboard would not involve any changes to the mathematics of either relativity theory or quantum theory.

John Bell was aware of this loophole when he told Paul Davies, "one of the ways of understanding this business is to say that the world is superdeterministic." Still, Bell and others of his community could not take superdeterminism seriously for they believed it collides with human free will. As we saw in compatibilism in the chapter on free will,

Bell need not had worried about free will's stance in a superdeterministic world. So his objection to the loophole in his inequality was more of a personal opinion than a professional statement.

For at least sixty years then, we have been misled into thinking theories of hidden logic or mechanics are inherently flawed and not to be explored. For hidden variable and local theories, all of this corrected information is overdue. The erroneous thinking has been deeply ingrained into our physics community. It will be hard to get a whole community of scientists to come around, especially one that has been trained to think that Nature is not supposed to make sense. We will just have to do our best in exploring these issues, and see what happens.

11

*Reinterpreting A
Reinterpretation*

There is an unrenowned reinterpretation of quantum theory that when itself is reinterpreted supplies us with a superdeterministic quantum theory. Remarkably, this superdeterministic quantum theory remains strict to the usual quantum formalisms. It is merely a reinterpretation of a reinterpretation of the existing mathematics. The only change is in how we think about quantum theory, and, thus, where we will go with it.

The first step to the superdeterministic theory is the transactional interpretation, developed by the physicist John Cramer at the University of Washington. Cramer uses an superluminal nonlocal model. As he points out, the transactional interpretation is not an ad hoc accessory to quantum theory to make it more logically coherent but rather a reinterpretation of wave mechanics and electrodynamics found within quantum theory.

Recall the earlier discussion on the laws of physics being time reversal invariant, working symmetrically in either direction of time, and the example in the Maxwell-Lorenz equation. Cramer takes advantage of temporal symmetry and of the very same notion of retarded waves and advances waves then couples them into a transactional model. To illustrate his method, let us go into more detail of the famous wavefunction. A wavefunction is written out as a complex number, meaning it consists of a real number plus or minus an imaginary number. When we want to calculate the probability of how a wavefunction is going to collapse (or with what numbers the dice will stop on), we multiply the complex number of the wavefunction by another complex number with an opposite sign in front of the imaginary part of the number. This is the standard method found in the Schrödinger-Dirac equation in quantum theory, and the operation is referred to as complex conjugation. For clarity, we can represent the wavefunction as (a—i). Given an initial quantum state for a particle, we can predict its future quantum state by squaring the wavefunction, which for complex numbers translates into: (a—i) x (a + i). Cramer breaks from orthodoxy by treating both of the complex numbers, (a—i) and (a + i), as wavefunctions each representing an event. Cramer describes one wavefunction as traveling forward in time and the other as traveling backward. In essence, he depicts the collapse of the wavefunction as the transaction between two quantum waves, one from the future and one from the past.

The attraction is in how this alternative interpretation handles quantum weirdness. It works like this: When an electron vibrates, it produces a field consisting of a wave traveling backward in time and a wave traveling forward, that is; the advanced wave and retarded wave, respectively. The forward retarded wave encounters an absorber electron which also vibrates, producing its own field and its own waves. The second electron's retarded wave exactly cancels the first electron's retarded wave so there is no further future effect. But the absorber electron's advanced wave heads back in time to the emitter electron making it recoil in a way that it produces another advance wave which cancels its first advanced wave. The emitter's retarded wave and the absorber's advanced wave then reinforce one another. What is left is a "handshake" between an "offer" wave and a "confirmation" wave.

If we return to Wheeler's delayed choice experiment and apply the transactional interpretation, we can see how Cramer's wave theory dissolves a paradox. The experiment verified that light can behave as if it knows ahead of time what will happen during the experiment. It will go through both slits and interfere with itself if it will later reach the film plate before any other detection. Or, it will pick only one slit if its path will be checked at sometime in the future. As Bell's inequality indicated, this seems to be evidence against either locality or independence.

If we study Cramer's transactional theory, however, we will notice it monopolizes on the loophole in Bell's inequality. It questions the assumptions of independence and finds where causation is temporally bi-directional, working as much from the future as from the past. Where as the delayed choice experiment shows precognitive light or spooky action, Cramer clarifies that this is only so when we assume causation to function solely in the direction of the retarded wave. In the transactional model with waves sniffing ahead and reporting back in time, it does not matter when detectors are turned on since the detectors are placed between the emitter and absorber which affect each other superluminally.

Cramer takes the first step in providing for advanced action, but his transactional model stops there. It is important to observe that the transactional interpretation describes two wavefunctions traveling in opposite directions in time, treating advanced waves as physically distinct from retarded waves, meaning they both correlate with real events. Yet, this distinction between the two waves is not accounted for in the building arguments. It arises straight from the assumption of time's arrow for which Cramer wants his model to abide by.

In other words, the transactional interpretation upholds nonlocality within the context of flowing time and is consequently rot with logical problems. For instance, he overly describes events as happening through time, forward and backward, to the extent where he has created a clear-cut case of a time paradox. Let us recreate the story Cramer's model tells, this time focusing on the emitter's advanced wave. At time 1, an electron A emits an advanced wave and a retarded wave. At time 2, an affected electron B emits an advanced wave and a retarded wave. At time 1, again, the electron recoils from an advanced wave causing it to produce another advanced wave. So which is it? Does electron A emit two advanced waves at time 1, or are there somehow two occurrences both labeled time 1? When a wave is produced and canceled out before it was actually produced, was the wave ever produced? Then there is the paradox with a wave causing another wave which goes back in time to reinforce the "first" wave; It seems the "first" wave would already be reinforced, but this sort of logic takes us on a loop, reinforcing the reinforced wave, ad infinitum. It is also unclear which electron is actually the absorber and which is the emitter. As it stands, Cramer's model seems to replace the usual paradoxes with new ones.

Alas, not all is lost. The transactional model is a good first step. We need only to follow up with something like Ockham's Razor, also known as Occam's Razor or the principle of parsimony, stating that (1) entities should not be multiplied needlessly, (2) the simplest of two or more competing theories is preferable, and (3) an explanation for unknown phenomena should first be attempted in terms of what is already known.

Ockham's Razor is precisely the type of scientific conservatism pushed throughout this book. It is time to apply it. If we take the transactional model and cut out the obscure notion of an arrow of time, we are left with a viable model. Instead of having the double-talk involved in keeping retarded and advanced actions distinct with different events overlapping in time, we simply look at calculations for advanced waves and calculations for retarded waves as two different ways of seeing the same sort of thing. The model can be seen as depicting how the mathematical formulas get at the answers without perfectly correlating with objective reality. From nowhen, the virtual photon being exchanged between electrons can be seen as expanded along the spatial curve of time where light can be seen as neither emitted nor absorbed but as a thread connecting two observables. From this atemporal perspective, the illustration on the right in the above graph is what is most real. Everything else is a symbolic leading-up-to process for arriving at the answer.

Now we have a superdeterministic model of classical proportions. It remains strict to quantum formalisms and is experimentally identical to the standard interpretation, including the original Copenhagen Interpretation as well as the recent multiple-worlds and many-histories interpretations. The only important difference is in the metaphysical luggage, that is; the superdeterministic model of quantum theory travels light. It is the kind of theory whereby you can tell a straightforward story about the world, the kind of story that contains nothing cryptic or unintelligible and nothing inherently ambiguous and in which no event is statistically determined, and all of which we are permitted to understand.

Those of us who speculate on modern physics while presuming there is an arrow of time resemble the land surveyors who believe land is flat. Earth only looks flat from where they are standing. We have reached a point in our discoveries where our instruments of increasing precision are recording inexplicable data. And since we as a scientific community have yet to look beyond time's arrow, we have failed to see the curve of time and fully appreciate how spatialized time and relational theories answer our puzzling experimental results.

12

Refraction Revisited

Quantum theory is abstract, its mysterious behavior distant from our everyday life. It is not difficult to dismiss its quirkiness as an artifact of symbolic reasoning. Sure, quantum theory has brought us nuclear power, lasers, microchips, PET scans, and other marvelous technologies, but even these are not understood by many of us. Conceivably, the problems of logic were created out of too many hierarchical assumptions and we have built ourselves into a tangle of mathematical loops. Perhaps.

Maybe another striking example is in order, one not entrenched in the intangible world of quantum mechanics but from our everyday world. Everyone has seen its occurrence, so often that we no longer inquire as to how it happens. Yet, if you ask a person on the street to explain why this everyday occurrence happens, you will not get a straight answer.

The phenomenon is refraction. You see it in the glass of water with the bent pencil. When taking a bath you notice your legs appear to be broken. Spear fishers aim below their targets as compensation. Eye glasses, cameras, telescopes, and microscopes all utilize this effect of refraction. It is something we grow-up with and accept as normal. But before diving into a pool, do you ever pause to wonder how light bends like that?

In the Middle Ages, the man known as Alhazen published the first geometrical descriptions of light in his *Opticae* thesaurus (*Treasury of Optics*). Let us recount his brilliant reasoning in the section concerning refraction. If we assume three propositions, then we can account for refraction. The assumptions, none of which were popular around the year 1000ad, are: (1) the velocity of light is finite, (2) light travels from a light source, such as the sun, bounces off of an object, then enters the eye, and (3) light follows the path of least time which seems to be a straight line. Now we need only add these propositions with the observation of refraction to see what we get. Output: light travels slower in water and glass than it does in air.

For clarity, an illustration of refraction follows. You see, light travels the quickest path between two points, and when one of those points is above the water's surface, say your eye, and the other is submerged, say the tip of a pencil, the quickest path is not a straight run through Euclidean space. This is because light takes longer to travel through water than through air. So the light from your submerged hand angles up a little to escape the water sooner, then hightails it to your eye, cleverly taking the overall path of least resistance and least time.

Light does this trick of cutting its travel time, but how? How does it know before arriving at your eye that a straight shot is not the fastest or the easiest? Thanks to Feynman's ingenuity, students of today learn to talk about the path of light as a tidy summation of an infinite number of possible paths. It is the quantum electrodynamics approach called path integration.

Feynman's trick was to frame the problem in purely abstract terms by building a mathematical model. We can crudely understand his trick by imagining a sheet of paper with two ink dots, one representing a starting point and the other being the ending point. Now our task is to go crazy with the pen, drawing connections from the starting point to the ending point, using curves and squiggles, any path we can think to draw, path after path until the entire page is blacked out. Next we assign values to represent each path. We then add the values, summing-over an infinite number, allowing all mirroring paths to cancel each other, and represent this result with the remaining path connecting the points. The sum total turns out to be a near straight line. This works for calculating refraction, as well, since the values representing light under water will be shifted in comparison to those above water, leaving a summed-over kinked path.

Is this hocus-pocus math or does light actually behave as the model suggest?

If we follow intuition in presuming time indeed flows, then path integration is a conceivable model. It holds for being a useful theory in predicting how light will behave. It has merit in its applicability. We can sit on the side of the pool without fear of precognitive light sniffing out paths of least resistance. It also predicts the rainbow of colors we see in the reflections of CD's. Hold a CD up to a light source, find the source's reflection in the CD, then tilt the CD away so that the source is no longer in the reflection, and you will see a colorful array of light from that source continuing to reach your eye along peculiar paths. The reason you can see them, as well as why they are colored, has to do with the spacing between the grooves and the wavelength of light. Remarkably, some waves which would normally cancel other waves are captured by the CD, allowing the other waves to register on your eye.

What if the presumed arrow of time does not exist? All things considered, science lacks substantial reason for treating the arrow of time as anything more than a provisional insertion into the theories of physics. Until time's arrow is vindicated by something more concrete than the present meditations, the notions of flying arrows and flowing rivers should remain outside of science. The arrow in the path integration model being no exception, how does the model stands without the dubious postulate?

Without time's arrow governing the theories of physics, Feynman's model retreats from describing underlying mechanics. It stands as a predictive model where its elaborate mathematics need not correlate with how light foresees the path ahead of least resistance. Light appears to cheat time so Feynman found a way to cheat as well, but this does not mean the two tricks are the same. The reality of how light behaves might be less of trick and more of a misconception on our part. Where Feynman describes a single photon as traveling an infinite number of paths so that the right one remains, the real photon might not be traveling any path. Asking how the photon knew the best way from point A to point B, may be as odd as asking the railroad track how it knew the best route through the mountains. If neither one is actually traveling, then we are looking at the problem the wrong way. In the case of a photon, what we should then do is look at its behavior from many directions at once, from its perspective, from ours, from future to past, and from nowhen.

In explaining phenomena like refraction, the secret is not in light traveling every-which-way. The secret is in how it does not travel. If we could ask a photon how it got to where ever it is, the photon would look at us like we were crazy. Indeed, we should be worried about our rationality. We acknowledge the theory of relativity but continue to see our passage of time as specially real. The photon is at point A and point B without any time passing for that photon. We look at this and claim time did pass. We do not care about the talk of frames of reference and time and space being intimately connected. We felt time pass. But we are wrong to discriminate this way. Experiments have confirmed relativity. Mechanical clocks which do not discriminate have been accelerate into orbit and brought back to show how genuine the connection is between time and space. If there was no change in time for the photon, then the photon is at point A and point B at the same time, and this perspective and description of events is as real as ours would be from our tangible frame of reference.

Here we have it; We are back to talking superdeterministically in physics. Whether we follow the logical conclusions from relativity or quantum theory, we can tidy up with spatialized time.

13

And Entropy Too

The Austrian physicist, Ludwig Boltzmann, is remembered for his important contributions to the kinetic-molecular theory of gases. By investigating the relationship between the temperature and the energy distribution of molecules in a gas, he laid the foundations of statistical dynamics. Boltzmann and his contemporaries, notably, J. Loschmidt, E. Zermelo and E. Culverwell, further hoped to discover the mechanical underpinning of the second law of thermodynamics but without success. To this day, the second law of thermodynamics remains statistical in nature without an underlying mechanical model.

Entropy is a semi-vague quantity specifying the amount of disorder or randomness in a system. It indicates the degree to which a given quantity of thermal energy is available for doing useful work; the greater the entropy, the less available the energy. According to the second law of thermodynamics, during any process the change in entropy of a closed system is either zero or positive; thus the entropy of the universe as a whole tends toward a maximum.

As mentioned above, the usual explanation for the second law of thermodynamics is a statistical analysis of causation. The reasoning in why systems tend toward thermodynamic equilibrium is as follows:

There are more possible disordered states than there are ordered ones. Suppose a system begins in one of the few possible ordered states. As time goes by, the system will either remain in the same state or it will change. At a later time, it is more probable that the system will be in either the same state or a less ordered state since disordered states outnumber ordered ones. Thus, if a system in inequilibrium evolves, then entropy will tend to increase with time.

By associating a thermodynamic arrow of time with statistical dynamics, physicists are necessarily predicting events from past to future. They are not attempting formulas that predict the other way from future to past. As a result, their work begins with a temporal bias and cannot be applied to events in the order of future to past. For example, imagine a game of shells. A pea is placed under one of three shells by person #1 while person #2 is looking away. Now person #2 has a 1/3 chance of guessing the correct shell with the pea. But if you were to film the entire game and play it in reverse and if you were to watch many games in reverse, you would say the probability of person #1 picking the correct shell to be 100%. You see, statistical analysis only depicts relative frequencies of events following the arrow of time which we presuppose exists. Probabilities are all about using black boxes and correlations, ignoring inner workings such as why person #2 picked the particular shell that they picked. For statistical purposes, person #2's detailed reasons for acting are summed over so that person #2 is categorized as one in the set of all persons. In other words, the determinism which is so obvious in films played backward is completely overlooked in films played forward then treated as a sum over possible histories filled with chance and accidents.

Say you have a table top covered with dice all with the number five facing up, labeled state A. Then you knock over the table scattering the dice across the floor. When they stop rolling each one is again revealing the number five, labeled state B. Indeed, this phenomenon AB would be bizarre. But how bizarre would it be to see the dice gathering up from the floor, accelerating up in the air along with the table, ending with the table upright and every die showing the number five, phenomenon BA? It may seem strange but let me remind you that BA obeys elementary physics. Furthermore, following temporal events from future to past, it is mechanically inevitable that B will lead to A.

The argument for the thermodynamic arrow defeats the likelihood of phenomenon AB, not BA. Since the laws of motion are time reversal invariant, the statistical justification for a thermodynamic arrow can be turned around and made for the past temporal direction as well. Therefore, not only is it highly likely that entropy will increase, but it is also highly improbable, depending on the temporal direction we choose to point the thermodynamic arrow. Without changing the basic laws of nature, there is no way to eliminate this subjective bias in the second law of thermodynamics. Any appeal to exclusively one temporal direction must be strictly provisional and acknowledged as such.

Interestingly, when we fine tune entropy by measuring a system within great detail, entropy never changes. Only at levels of detail which we consider significant for the discussion of processes does entropy in a sense change over time. This is its statistical nature. Accordingly, when looking for a clue or proof for an irreversible thermodynamic arrow in a statistical theory such as entropy, we embarrassingly resemble the dog chasing its own tail. Where these theories are predictive, they lack the detail of real events, and when they try to bridge this gap they fall into assuming the very thing they are trying to prove, time and time again.

Physicists typically rely on this erroneous conceptualization of a thermodynamic arrow to account for an objective arrow of time. They say entropy increases in the same direction that time flows and that there is somehow a connection. Richard Feynman, in his lectures on The Character of Physical Law, told students, "The apparent irreversibility of nature does not come from the irreversibility of the fundamental physical laws; it comes from the characteristic that if you start with an ordered system, and have the irregularities of nature, the bouncing of molecules, then the thing goes one way" (1994, 107).

The second law of thermodynamics states that a closed system's entropy will either be constant or increasing. But there is a problem with this. The laws of physics are time symmetric and without this second law a physics textbook cannot tell you why glasses break but don't come together or why different fluids mix but do not separate when going from the past to the future. So when a physicist wants to be able to say that a movie film showing a shattering glass is being played forward they must rely exclusively on the second law to make this claim.

Stephen Hawking, in his best seller *A Brief History of Time*, commented, "Disorder increases with time because we measure time in the direction in which disorder increases" (1988, 147), and said so without recognizing how obviously non-profound is his statement. Here is the circular logic in its overt form: If time is measured from what we call the past to what we call the future, then the entropy in a closed system will either increase or remain constant. If the entropy increases in a closed system, then that system is moving from the past to the future. This is the absurd form: If X then Y, and if Y then X. This means we have a correlation between past/future and low/high entropy, but no cause-effect relationship can be made. Consequently, Feynman's tautology that Nature is irreversible because it goes one way, is not saying anything interesting. The second law of thermodynamics cannot be used in revealing an arrow of time because it already assumes that very same arrow. Physics has not an arrow of time, but a pattern in time. Thus, a more proper form of the second law of thermodynamics is: Entropy either remains constant over time or it increases in one direction of time and decreases in the other direction. As Huw Price puts it, "What is objective is that there is an entropy gradient over time, not that the universe 'moves' on this gradient in one direction rather than the other" (1997, 48).

If we do not treat spacetime in reference to some imagined external clock and instead simply think of spacetime alone, then there is as much reason to believe that entropy decreases toward the past as there is for believing the reverse is true. This is because the very same process is proof of both, not one or the other. To clarify, the statement, "the reverse is true," does not mean, "when time is reversed, the same is true." The latter creates two separate processes for every one process with each one oriented in opposite directions. The latter also supposes an arrow of time along side space instead of spacetime. The former only implies there is a single history where rules of causation are symmetric. A single process equally supplies evidence for entropy increasing in one direction of time and decreasing in the other direction. The existence of eggs is proof for broken eggs assembling themselves as much as scrambled eggs are proof of the opposite.

Nevertheless, significant entropy can be discussed while ignoring any arrows of time. This is done by spatializing time so that all states of the universe along the dimension of time co-exist at separate points in time. We can then compare the entropy between certain states, but in order to avoid logical errors we must compare these states without assuming any arrows, flows, or irreversible processes. Here's an analogy: Say you hold a measuring stick in your hands, and this stick happens to be painted like a rainbow with the spectrum from one end to the other. Say at .25 meters the stick is orange colored and at .5 meters the stick is yellow. Here we have a difference in color at length .5 and .25, but this difference neither assumes nor implies that color necessarily flows from either .5 or .25, especially since the stick is already painted. The same applies to co-existing, or spatialized, states of a system.

Over here. Follow me for a moment. Let us step outside of the universe, looking back on it from an imagined place in superspace. It is a difficult objective position but an enlightening perspective as well. Now, looking back, we see the universe in its full form of spacetime, all places, all times, up, down, left right, past, future. If we raise the question, "Why does significant entropy increase with time?" we see this is only true in one direction. The reverse is equally true in the other direction. From our special perspective, the arrow of time is better understood as not an arrow of time but a pattern in time. As pointed out earlier, the second law boils down to an observation of a pattern across time and that the universe in one area of this pattern, what we call the past, has a character of low significant entropy while in another area, what we call the future, has high significant entropy.

A group of international scientists stated in a recent press release, "The reason why the future and the past are so different in our daily lives is because the Universe started off in the Big Bang in a smooth and organized state...and is evolving into some disorganized state, simply because there are so many more disordered states. This is the origin of the large-scale arrow of time" (CERN, 98). You might hope for more coming from CERN. The writers made at least to loops of logic, one within the other, and though they are common mistakes, one should wonder how they passed for scientific publication. The loopy thinking, starting and ending in the same place, is a fun way of talking ourselves in circles. So fun, in fact, that I think we should try it, but let us do our own version concluding that the arrow of time flies toward the past. Here: "The reason why the past and the future are so different is because the Universe started off in the future in a rough and disorganized state and is evolving into an organized state, simply because there are so few tidy ways to be. The universe naturally tends to become more and more tidy. This is the origin of the arrow of time."

Crazy talk, perhaps, but only on the surface is it more wacky than anything else quoted in this book. Claiming that our sense of a moving present is more than vertigo, so much more that it parallels a cosmic river of time, unnecessarily convolutes the mind of the empiricist. Is this not the grandiosity of our place in the universe?

14

Cosmic Marbletop

A wise man of our times wrote, "Far out in the uncharted backwaters of the unfashionable end of the Western Spiral arm of the Galaxy lies a small unregarded yellow sun. /Orbiting this at a distance of roughly ninety-eight million miles is an utterly insignificant little blue-green planet whose ape-descended life forms are so amazingly primitive that they still think digital watches are a pretty neat idea." This wise man lives in London, of all places, and is the author of the indispensable *Hitchhiker's Guide to the Galaxy*, of all books. Douglas Adams' satirical words are an appropriate introduction to this chapter on cosmology and the structure of the universe because there is indeed truth to be had in fiction. Is it not peculiar that a group of about a thousand self-selected humans/cosmologists speak about the ultimate beginning and shape of the universe? Here we are in a remote corner of space, which extends at least seven billions light years from us in all directions, and we, who are under seven feet tall and argue over where is the best place to eat, actually have what we call 'theories' to explain the large -insanely huge- scale structure of the universe. Does this not sound a little odd? Within my lifetime, cosmology has gone from being a joked about pseudo-science to being a creditable branch of physics. The real kicker is that much of what cosmologists say in their characteristic matter-of-fact lectures is *mostly* true.

In spite of our reasons for feeling smug, this chapter would not had been written if I did not have a bone to pick with respectable cosmologists. Of course, my objection has to do with how physicists treat the concept of time. You see, however close the physics community comes to assessing what the arrow of time boils down to, it is invariably disappointing to find the flow of time seeping back into their assumptions. To confine the seemingly infinite measure of time, modern physics will either talk of a closed universe or they often seek out boundary conditions. Not surprisingly, the big boundary is the state of the universe in the first moment of the hypothesized Big Bang. This is called the initial boundary condition for the entire universe. Notice the usage of "initial." Here we have the flow of time leaking into one of the grandest areas in physics; cosmology.

But I am getting ahead of myself here. To properly explain in what way I think cosmologists are wrong, I will need to back up a century or two. So without further ado, allow me to digress.

Do we know infinity? What does it mean to say something is infinite? In usual Euclidean geometry, infinities exits within the limited geometry. On the other hand, in curved Riemann geometry, a line might become a closed circle, an infinitely extending plane might become a sphere, an infinite space might become a hypersphere, and so on. Infinities exists within imposed limits and can fade as we increase our level of perspective. We do not know if infinity is only a concept within the confines of our minds. It may be the case, to create an aphorism, infinities exist in the mind because the mind is not infinite.

Our everyday experience tells us that time is linear. So when we think of the evolution of the universe, we imagine it going from past to future. And when we think on this further we see a dilemma. Time must either be infinite or there must be a beginning and an end. Saying there was a beginning to the universe leads to the natural question of what was before the beginning. To say there was a creator merely introduces new terminology without solving the puzzle. This may remind you of the question, "How did life on Earth start?" and the empty answer, "It came from outerspace," followed by, "How did life in outerspace start?" and so on. If a creator created the universe, how was the creator created? We are still stuck. We could give up and say that time is infinite or that there was a beginning and nothing was before the beginning, or nothing created the creator who was before the beginning, or nothing created the creator who created the creator who was before the beginning, or. . . No matter how we put it, a temporal line boggles the mind with philosophical puzzles.

Some intellectuals, David Hume, Noam Chomsky, Gunther Stent, Colin McGinn, Jerry Fodor, and Steven Pinker, to name a few, have enjoyed being provocative by saying some answers are beyond human potential understanding. Pinker says, "We are organisms, not angels, and our minds are organs, not pipelines to the truth" (1997, 561). Whereas an ant cannot comprehend a combustion engine and a monkey cannot fathom the intricacy of a processor chip, nor can Homo sapiens understand the mystery of time. The philosopher Daniel Dennett has a neat reply to such a view of cognitive closer. "We should be unimpressed by the examples, for not only can the monkey not understand the answers about electrons, it can't understand the questions. The monkey isn't baffled, not even a little bit" (1996, 383). An ant cannot be baffled by a combustion engine. If they could be baffled and if they could pose the questions to themselves, then perhaps they would also be able to understand the answers. Our understanding of anything is linked to the kinds of questions we can think to ask. Is it not even peculiar that we ask ourselves of the very question of cognitive closer? How many other animals have wondered about their cognitive limits? Since we are quite aware of the enigmatic character of time, some sort of understanding is in order. The other option is to give up trying to make sense of time and in attempting to imagine what some say is impossible to imagine. Before opting for cognitive closer, let us see if the problem is really so difficult. After all, what is unthinkable for us is truly unthinkable and cannot be discussed in this book.

For fun, let us imagine a wonderful analogy in the world of Zargons. The people of Zargo cannot decide on the shape of their land. The philosophers believe land is infinite in all directions and infinitely thick below them. The priests believe land has a surrounding edge where they can fall off the end of the world and that the land has a bottom where, if they dug deep enough, they could fall through. The priests also believe their god, Zar, holds the land up so it does not itself fall. Some philosophers wonder what holds Zar up, while others argue that some things are too perplexing for Zargons to comprehend. Generations later, Zargo explorers learn that both groups of ancestors where wrong, that their land is neither infinite in area nor with edges to fall over. This is because they live on Zargon, a planet.

By combining the earlier works of Dante (who metaphorically described a hypersphere in *The Divine Comedy*), Riemann (who formalized the hypersphere in pure mathematics), and Einstein (who combined the metaphor with the formalism), an underground designer fashioned these holistic eye-glasses. Here, try them on. Do you see how a spatialized temporal dimension can be rapped up like the surface of a planet, with no beginning or end? Look how all of the dimensions can create the spacetime continuum with no edges or boundaries. Watch how time simply becomes North and South on this four-dimensional sphere. Do you see the feathered arrow of time anywhere? Of course not, all you find is a featherless compass needle pointing one way without any flight plans. It does not move. In fact, since all places and all times are mapped on the hypersphere, nothing moves. With these glasses on, you can see the events of yesterday and tomorrow and all that happened in-between presented along the hypersphere. Okay, now take off those funny looking glasses.

Imagine a grossly over-simplified illustration of the universe. Changes in time are shown by North and South. To your left is a circular mapping of the universe as it is in one moment of time. It looks like a circle. To your right is a spherical mapping of the entire universe in all moments of time. In a world with an arrow of time, there would only exist the circle to your left, which would evolve through time, from South to North. What is real is only the dynamic of the circle. On the other hand, in a world of spatialized time, the circle and the sphere are equally real, picking a time is like picking a place, or one of an infinite number of circles within the sphere. "Later on" is as real as "over there." Notice how we can roughly achieve one from the other. The dynamic circle can be replaced by tracing the dynamics through a continuum of non-dynamic circles. Vise versa, we can derive the circle mechanics by deconstructing the non-mechanical spherical model into an infinite number of circles. At this point, we should ask ourselves which model is fundamental to the other. Which is a more accurate description of our universe? Or, even better, which model best fits observations and best predicts future observations?

In answering these questions, we have already seen in previous chapters the problems posed by the dynamic model. The most obvious is in how time is treated as fundamentally dynamic. There is not a more basic model to provide an explanation for the over-all process. This complicates the theory of relativity and quantum theory because a flow of time must be inserted into the equations as some sort of truism, resulting in some frames of reference being more special than others and, in turn, quantum weirdness.

The non-dynamic spherical model is basic to and can derive the circular model with its arrow of time. It can also derive the circular model's counter-part which travels in reversed time. One dynamic model represents retarded mechanics and the other shows advanced mechanics. Where one fails to account for data, the other steps in and together they present the solution. The Schrödinger-Dirac equation is a prime example. Thus, the spherical model encompasses all the laws of motion.

Quick note on Before Time: Can time have a beginning? Where would it be, and what would be before it? Does the surface of our planet have a starting point? "Oh, you're looking for the beginning of the earth? Well, head *east* until you reach it." The surface of the earth has no beginning nor starting point. For this same reason, it is illogical to claim that the Big Bang is the beginning of time. One might as well say that right now, or tomorrow, is the beginning of time. One can still imagine a point outside of spacetime, like at satellite, but imagining it does not make it real. Computers can manipulate objects in n-dimensions, but that doesn't mean there are 11, 22, 4862348, 2/3, or whatever-you-want-n-to-be-dimensions (possibly but not necessarily). As far as we know, the laws of physics/cosmology in no way suggest an uncaused first cause or creator, nor final cause, for that matter.

15

The Infamous Kaon

Not mentioning any names, it has been noticed by this author how other authors of similar philosophy show little respect for the unique decays of the kaon and its antiparticle. The peculiar kaon pair when mentioned at all is often mentioned in passing as being poorly understood and inconclusive, which is true, but by giving it such little airtime, it is left as a nagging needle in the side of all superdeterminists. A dismissive attitude wins no supporters. That is why this author has devoted an entire chapter to the infamous kaon.

You may remember way back to the first chapter the allusion to the nonstatistical laws of nature being time reversal invariant, meaning that they are indiscriminate between the past and future directions of time, working equally well either way. This is apparent in mechanical theories such as in the Schrödinger-wave equation and Newtonian laws of motion. In contrast, there are macro and micro laws of mechanics which are not time reversal invariant because they follow aggregate patterns in physics. So while billiard balls and photon waves are indistinguishable in positive or negative time, the aggregate behavior found in radiating bodies and mixing gases are undeniably biased in their directions of time.

On the micro-level, besides radiation, decay, and other quantum phenomena exhibiting temporal bias, the physics community focuses on a special case of temporal asymmetry found in kaon decay. Those who are looking for an inherently fundamental arrow of time will make reference to this decay, implying that the asymmetry is evidence of flowing time.

After reading Chapter 7 on probability theory, the reality of kaon decay reveals itself to be less profound, that is; after learning the asymmetry is strictly a statistical one. Kaons decay slightly less often into anti-kaons, or "mirror" kaons, than anti-kaons decay into kaons. On an average of about one in a thousand, one line of decay happens more frequently than the other, and since "mirror" particles are supposedly temporally symmetrical to one another, the difference in averages hints at temporal asymmetry, though it should be noted the decay can lead to at least one conclusion other than temporal asymmetry. Seeing as we are talking about a mirror kaon, it may be the case that there is a spatial asymmetry between left and right. Or, with respect to relativity, there could be a temporal/spatial asymmetry. Why we tend to focus on the temporal seems to be a matter of convention.

Furthermore, finding a trace difference shows an overall pattern but has no implications for the reversibility of a single decay or event. The reason is two-part. One, as shown in previous chapters, it would be a mistake of logic to use probability theory in explaining single events. Since the given anomaly appears only in statistical results, we are unable to extrapolate the numbers for dealing with explanations of "how it is that way" for any lone kaon. The other reason, not mentioned in any previous chapters, is that symmetry and reversibility are not interchangeable, nor do they necessarily go hand and hand. If an experiment shows a process to be temporally asymmetrical and a dominant theory is unable to provide for the bias, this only means for certain that the theory is incomplete. It would remain to be seen whether or not the process is reversible.

One last point specific to the kaon: Given that this particle is a rare and exotic form of matter (and it is far from clear how an elusive particle's behavior might be connected to the rest of the world), specialists and related physicists alike do not pretend to understand the relevance of kaon decay to general phenomena. What the asymmetry means for physics as a whole is anyone's guess right now.

Getting back to statistical vs. reversible theory, if all we had in science were the reversible laws of physics, we could not deduce the world as we know it. Boundaries or conditions are essential to getting the right answers. These additions to theory have taken their practical form of probabilities. Professor of chemistry and physics Ilya Prigogine and his group at the Ilya Prigogine Center for Studies in Statistical Mechanics and Complex Systems have argued that we should stress the probabilistic side of physics and view the typical nonstatistical side as less basic to understanding the universe. By their light, temporal asymmetry of phenomena are seen to arise from nonlinear dynamics fundamental to our physics. In three words, asymmetry from asymmetry.

Prigogine and his group have put forth that the laws of statistical mechanics are the basic laws of physics. To recount what is meant by "basic," let us reference the old hierarchy working behind the scenes in today's science. The hierarchy is based on a pyramid with degrees of fundamental science working from the ground on up, with Mathematics and Physics along the bottom, then Chemistry, Paleontology, and Biology and so forth, on up to the softer sciences. It assumes the reductionist's ultimate stance; where everything can eventually be explained by lower levels of study. So what the Prigogine camp's claim amounts to is nothing less than putting probability theory at the very base of all of scientific theory.

Sounding familiar? It is the cry of a current paradigm shift; probability all the way down; super*in*determinism, as I call it. Given that it postulates the world to be probability to the core, it accordingly posits mathematics as the foundation, which is the epitome of a modern

Platonist's sublime dream where all the world is envisioned as a pedigree of pure mathematics.

Immediately we notice a problem with the above picture. We can not help but wonder how any one thing can be real when everything is made not only mathematical but statistical nonetheless. By basing reality on a mathematics which prescribes rules exclusively to *aggregate* patterns, we are left with no *single* reality. Do summed-over events then become nonexistent? If so, then how did these events get to be summed-over in the first place? More importantly, who or what is doing the summing-over? Probabilities are calculated by choosing handles or boundary conditions or something limiting, so who or what makes these decisions then calculates the odds? Without these handles, indeed without an associated perspective, there are no odds. It would be a mistake to answer that the universe is like a calculator because this lays a mechanical model beneath the indeterminism, being that a calculator is a mechanical device. So we might ask how the probabilistic equations calculate themselves. We also question the above paradigm's answer to the end-to-all-question, "How could it be that way?" We run smack into: "It just is!"

If we had to, we could live with the inscrutable super*in*determinism. Fortunately the more classical notions of physics continue to work within the modern framework of science. Nothing so utterly abstract is required for our ontology of the world. Super*in*determinism, with its hefty metaphysical baggage, is an alternative metaphysics being explored by prominent philosophers and scientists, but, in the meantime, a simpler and more conservative picture will be put forth by the author of this book, a picture which has and will come together and will hopefully be clear by the final pages.

16

One More Time

Okay, so it has been said what quantum theory, and more generally probability theory, does not mean for reality. At the same time, no one denies just how well probabilistic theories do work. They gives us incredibly accurate answers, verified to many decimal places. It was also mentioned some chapters back how nonprobabilistic theories alone lack the ability to model our complex world. These reversible theories cannot tell us which way in time a process has occurred. What does this mean? Simply put, our theories are incomplete. Examples abound: The quantum retreats from being smooth and geometrical while relativity stubbornly refuses to be packaged into discrete units. Macro and micro physics have been at odds from the start. One set of mathematics and philosophy can hardly be translated into the other. Aside from the mathematics, amazingly enough, there is a tidy philosophy in which the above conflicts can peacefully coexist. It is an old one, been around for who knows how long or how many times it has appeared. In distinguishing it in its most progressed form, we call it superdeterminism. Granted, it does not solve mathematical differences, but, as has been demonstrated throughout previous chapters, it sure does put these differences in their proper place, as being mathematical in origin and not philosophical or metaphysical.

The alternative is hardly scientific. (Post)Modern science affords a liberating space, a place of uncertainty where a comprehensive understanding is not essential to being a good theorist and where modern narratives are not restricted to realistic accounts, a place where the world is able to dispense rational explanations. To paraphrase the contemporary fable:

"The truly indeterminate nature of unmeasured physical properties has to be accepted, and we should concern ourselves only with what is measured and ignore resolutely all else. We do not know all there is to know about the subatomic world, but we are certain of its uncertainty. Do not worry about this either, because the equations get us the right answers, never mind how."

The essential ingredients of the standard interpretation of modern theory include a pound of proof against locality, two cups of proof against hidden mechanics, a dash of uncertainty, then mixed with an unobservable wavefunction and poured over a base of poorly understood probability theory. The studying physicists is fed the ingredients and overwhelmed by the mostly subtle speculations and misconstrued hypotheses. Even the instrumentalists will often fall into the mainstream interpretation by declaring, as Hawking did, that indeterminacy is a fundamental, inescapable property of the world. Note the certainty and conviction in this statement. Yet, as was shown throughout this book, the proofs are disproved, the uncertainty is classical, the unobserved wavefunction is abstract, and probability theory is statistically predictive, not explanatory for individual events.

In investigating the principles of Nature, let us keep in mind that quantum theory is a patchwork mathematics. It is incomplete yet has been continuously molded and sharpened over many decades and generations of physicists by use of ingenious tactics so it would fit mountains of otherwise inexplicable data. Typically, quantum theory hits a wall, then a genius comes along figuring out how to renormalize the equations and be rid of unwanted impossible answers, and so here we are with a fairly precise formalism. Of course it is accurate. This was the entire goal, but unfortunately Platonistic physicists have been reading too much into the mathematics, and now we are left with a universe which is uncertain of itself. The thing is, it remains absurd for a scientist to say a universe is uncertain. Bohr was over zealous in borrowing from psychology (i.e.; William James' "complementarity") and from mysticism (i.e.; Eastern thought's "duality") and his colleagues were wrong to accept the needless obscurities, perpetuating the bamboozle. Generations of modern Platonists have tried to legitimize incomprehension into a science, however, "duality" and "uncertainty" are lame terms to have in physics. They belong in pseudo-science, spirituality, and mysticism. The physical world is not ambivalent about its state of existence. We humans are the ones who find it difficult to let go of our precious terminology of the macroworld, such as "waves" and "particles," so we chase phantom wavefunctions and see a ghostly world that is unable to make up its mind of whether it is here or there. Please, let us not assume that indeterminacy is anything more than the result of not being omniscient entities with divine knowledge, especially when quantum theory has been shown to be logically determinate by other conceptions of the very same mathematics.

To accept something like dualism, where some things operate by omni-mysterious stuff, like pure information, is to go beyond sober science. Such dualism runs into the problem of explaining how something like pure information can exist or affect the material world without itself being part of the material world. It has no counterpart in physical reality. Information is never pure. Information is hierarchical, based on the configuration of physical phenomena. Thus, there is a hefty amount of metaphysical baggage to carry if we deny superdeterminism and embrace dualism. If we are to deny the ontology of determinism, we should wait until there is ample reason for doing so, and thus far there is not, certainly not in quantum theory, this century's best evidence for advanced action and, therefore, welcomed in determinism. Indeed, quantum weirdness is the cherry on top when it comes to the history of philosophy and science in arguing for superdeterminism.

Superdeterminism helps cut through miracles and spiritual aspects of reality. It is the ultimate stance for any critic. The unexplained is, in principle, the yet to be explained. Superdeterminism will motivate the natural philosopher in their search for underlying logic in the world, especially in the instances when it is not obvious or appears counter-intuitive. Everything is seen to have a reason behind it. Much of modern dynamics is then viewed as operating with incomplete models while progressing in some of the most intellectually challenging fields of research.

The modern determinism put forth by this book reveals a consistency between the old and new which is greater than previously thought. My deliberate stance has been that classical physics can be reasonably advanced to account for today's scientific knowledge. Though this stance is counter to the consensus in the science of the last century, which has been heedlessly shifted from classical determinism to modern indeterminism, I hope my arguments have been persuasive enough to compel readers to rethink the predicament in today's theoretical physics.

Bibliography

CERN. "Time's Arrow: Particles Can Not Go Back To The Future." *http://press.web.cern.ch/Press/Releases98/PR06.98ETime'sarrow.html*. June 98.

Davies, Paul. *About Time.* New York: Touchstone, 1996.

Davies, Paul and John Gribbin. *The Matter Myth.* New York: Touchstone, 1992.

Dennett, Daniel. *Darwin's Dangerous Idea.* New York: Touchstone, 1996.

Feynman, Richard. *The Character Of Physical Law.* New York: Modern Library, 1994.

Geldard, F. A., and C. E. Sherrick. "The Cutaneous 'Rabbit,'" *Science*, 178, 1972.

Hawking, Stephen W. *A Brief History Of Time.* New York: Bantam Books, 1988.

Kolers, P. A., and M. von Grunau. "Shape And Color In Apparent Motion," *Vision Research*, 16, 1976.

Kosko, Bart. *Fuzzy Thinking.* New York: Hyperion, 1993.

Lindsey, David. *Cosmology And Particle Physics.* New York: American Association Of Physics Teachers, 1991.

Pinker, Steven. *How The Mind Works*. New York: W. W. Norton & Company, 1997.

Price, Huw. *Time's Arrow*. New York: Oxford University Press, 1997.

Savant, Marilyn vos. *The Power Of Logical Thinking*. New York: St. Martin's Press, 1996.

Smolin, Lee. *The Life Of The Cosmos*. New York: Oxford University Press, 1997.

Weinberg, Steven. *Dreams Of A Final Theory*. New York: Pantheon Books, 1993.

Printed in the United States
125410LV00002B/78/A

9 780595 178742